The Teen's Guide to Saving the Planet

Practical Steps for a Sustainable and Bright Future

Copyright © 2024 by A. Peters

All rights reserved. No part of this publication may be reproduced, stored or transmitted in any form or by any means, electronic, mechanical, photocopying, recording, scanning, or otherwise without written permission from the publisher. It is illegal to copy this book, post it to a website, or distribute it by any other means without permission.

Disclaimer: This book is a work of non-fiction and is intended solely for informational and educational purposes. The names mentioned within are trademarks of their respective owners. This publication is not affiliated with, endorsed by, or sponsored by any of these trademark holders. The inclusion of these names is meant to provide context and historical reference.

The author does not claim ownership of any trademarks or copyrights related to the names and likenesses of the individuals referenced in this book. Any opinions expressed herein are those of the author and do not necessarily reflect the views of any organization or trademark holder.

First edition

INTRODUCTION	1
UNDERSTANDING CLIMATE CHANGE	5
REDUCING WASTE	17
SUSTAINABLE EATING	26
ENERGY EFFICIENCY	35
ECO-FRIENDLY TRANSPORTATION	45
WATER CONSERVATION	52
SUPPORTING SUSTAINABLE BRANDS	58
GARDENING FOR THE PLANET	69
ADVOCATING FOR CHANGE	75
CLIMATE JUSTICE	80
ECO-FRIENDLY TECHNOLOGY	91
SUSTAINABLE FASHION	96
MINDFULNESS AND SUSTAINABILITY	105
COMMUNITY INVOLVEMENT	112
THE FUTURE OF OUR PLANET	123
FINAL THOUGHTS	133

Chapter 1

Introduction

Time's ticking, folks. The reality of climate change isn't some far-off threat anymore; it's knocking on our door, demanding we take action—like, yesterday. The ice caps are melting faster than ever, extreme weather events are becoming more frequent, and the delicate balance of our ecosystems is being disrupted. As the next generation, you've got the power to change the game. This guide? It's your playbook for making a real difference.

Picture this: a world bursting with life, where clean air and water are standard, and living sustainably is just how we roll. Imagine breathing fresh, crisp air, drinking water so pure it sparkles, and seeing thriving ecosystems teeming with diverse plant and animal life. This isn't some fairy tale; it's totally doable if we all pitch in. Each chapter here is packed with the know-how and steps you need to help create a greener, brighter future.

First up, you have to get a grip on climate change. Dive into the science, and you'll see how your daily choices fit into the bigger picture. Understanding the greenhouse effect, carbon footprints, and the interconnectedness of Earth's systems is crucial. Everything from your meals to your

Introduction

shopping habits can either mess things up or help heal our planet. So, let's own it—change kicks off with you.

As you flip through these pages, you'll find cool ways to cut down on waste and lighten your environmental load. You'll learn to make smart choices, like going plant-based or supporting eco-friendly brands. Discover the impact of fast fashion and the benefits of reducing meat consumption. Get ready to be inspired by creative hacks for upcycling and reducing plastic waste, all while championing local businesses that care about sustainability. Support your local farmers market instead of large supermarket chains.

Now, let's talk energy and water. These aren't just trendy terms; they're crucial for keeping our planet healthy. You'll discover easy changes you can make at home and the perks of using renewable energy. Learn about switching to LED light bulbs, unplugging electronics when not in use, and the benefits of solar and wind power. Plus, understanding the global water crisis will show you how everything's connected. From water scarcity in certain regions to the pollution of our oceans, every drop counts.

Transportation? Big deal. You'll learn how different ways of getting around impact the environment. Explore the carbon footprint of cars, airplanes, and public transport. Whether you're biking, walking, or pushing for better public transport, your choices matter. Together, we can build communities that focus on sustainable transport and cut down on fossil fuel use. Imagine cities with more bike lanes, pedestrian walkways, and efficient public transportation systems.

But wait, there's more. You'll get tips on developing a sustainable mindset, using mindfulness to live more eco-consciously. Learn to appreciate nature and the interconnectedness of life. Connecting with your community and jumping into local sustainability projects will show you just how powerful collective action can be. Join local cleanups, support community gardens, and get involved in environmental advocacy groups. Youth activism is a force to be reckoned with, and your voice can make waves far beyond your backyard.

As we tackle climate justice, you'll see how environmental issues tie into social equity. Understand how marginalized communities are disproportionately affected by pollution, lack of access to clean resources, and the impacts of climate change. Climate change hits marginalized communities hardest, and it's on us to stand up for those who bear the brunt. Let's back initiatives that promote climate justice and make sure everyone gets a piece of the sustainable pie.

In this tech-driven world, innovation can be a game-changer for sustainability. Explore advancements in renewable energy, sustainable agriculture, and waste management. From eco-friendly gadgets to sustainable fashion, our choices can either add to the problem or help solve it. This guide will help you think critically about what you buy and build a wardrobe that shows your love for the planet. Learn about sustainable materials, ethical production practices, and the impact of consumer choices.

So, as you set off on this journey, keep this in mind: the future of our planet is in your hands. Imagine a world where sustainability isn't just a fad but a lifestyle. With knowledge, awareness, and actionable steps, you can be a key player in creating a brighter future for those who come after us.

Chapter 2

Understanding Climate Change

Ah, the weather. When I was a kid, I thought it was just that—weather. Rain meant a day indoors, the rhythmic drumming on the windowpane a comforting sound; snow meant hot cocoa, the warm mug cradled in my hands as I watched the flakes drift down; and sunny days were for riding bikes until the streetlights flickered on, casting long shadows across the neighborhood. Simple, right? But oh, how naive I was! Fast forward to today, and I realize that weather isn't just a random occurrence; it's a complex interplay of science, human behavior, and, let's be honest, a whole lot of trouble if we don't get our act together. So, let's break it down, shall we? We're diving into the nitty-gritty of climate change, exploring the science behind it, how our daily lives are adding fuel to the fire, and why your individual actions matter more than you might think.

Let's start with the science behind climate change. Now, I'm not saying you need to be a meteorologist or a climate scientist to get this, but a little understanding goes a long way. Picture Earth as a big ol' greenhouse. Sunlight streams in, warming the planet, and some of that heat is trapped by gases in the atmosphere. It's cozy in here, right? That's because of greenhouse gases—like carbon dioxide (CO_2)

and methane—that trap heat from the sun. It's like wrapping yourself in a warm blanket. But here's the kicker: we're piling on those blankets like it's a winter storm! We're adding more and more greenhouse gases to the atmosphere, trapping more heat than ever before.

According to NASA, CO_2 levels in our atmosphere are higher than they've been in 800,000 years. Scientists have analyzed ice cores, which act like time capsules of the atmosphere, to determine these levels. That's not just a number; it's a wake-up call! Human activities—especially burning fossil fuels for energy, deforestation, and industrial processes—are cranking up the heat. When we burn coal, oil, and natural gas, we release CO_2. These fossil fuels, formed over millions of years, release stored carbon into the atmosphere when burned. And when we chop down trees, we lose a key player in the game: trees absorb CO_2 through photosynthesis, acting as natural carbon sinks. It's like throwing a party and forgetting to invite the bouncer who keeps things in check.

Now, you might be thinking, "Okay, but what's the big deal?" Well, here's where it gets real. Climate change leads to extreme weather events—think hurricanes, wildfires, and droughts. The increased heat fuels stronger storms, drier conditions that lead to more intense wildfires, and prolonged periods of drought. It messes with our food supply, causing crop failures and impacting agricultural yields, causes sea levels to rise, threatening coastal communities and ecosystems, and can even lead to species extinction as habitats are destroyed and ecosystems are disrupted. In short, it's a hot mess (pun intended) that affects all of us.

But hold on, let's not throw our hands up in despair just yet. Yes, the situation is serious, but understanding it is the first step toward action. And speaking of action, let's dive into how human activities contribute to this global warming catastrophe.

We've established that burning fossil fuels is a major player in the climate change game. But let's break it down even further. Transportation, for example, is a huge contributor. Cars, trucks, planes—you name it—emit a ton of CO_2. Every time an engine burns gasoline or diesel, it releases greenhouse gases into the atmosphere. In fact, according to the Environmental Protection Agency (EPA), transportation accounts for nearly 29% of total greenhouse gas emissions in the U.S. That's a staggering number! So, next time you hop in the car for a quick trip to the store, ask yourself: could I walk, bike, or take public transit instead? Even carpooling with a friend makes a difference.

Then there's our beloved agriculture. You know that juicy burger you had last night? Well, raising livestock produces methane, a greenhouse gas that's over 25 times more potent than CO_2 in the short term. Cows, sheep, and other ruminant animals release methane during their digestive process. So, while we're all about that farm-to-table life, let's not forget that our food choices have an impact too.

And let's not overlook waste. Landfills are like ticking time bombs of greenhouse gases. When organic waste decomposes, it releases methane. Food scraps, yard waste, and other organic materials break down in landfills, producing methane as a byproduct. So, when you toss that half-eaten sandwich in the trash, you're not just wasting food; you're contributing to the problem. Composting your food scraps is a great way to reduce this impact.

Now, you might be sitting there, feeling a bit overwhelmed. But here's the good news: individual actions matter. Yes, YOU can make a difference! It's like that old saying, "Think globally, act locally." It's not just a catchy phrase; it's a call to arms.

You don't have to go full-on eco-warrior overnight, but small changes can lead to big results. Start with simple steps like reducing your energy consumption. Turn off lights when you leave a room, unplug devices you're not using, or switch to energy-efficient appliances. These actions might seem minor, but they add up. In fact, the U.S. Department of Energy estimates that by using energy-efficient products, you could save about 25% on your energy bills. That's cash in your pocket and a step toward a healthier planet!

Then there's the power of recycling. It's like giving new life to old stuff. Paper, plastic, glass, and metal can all be recycled and turned into new products. Did you know that recycling one ton of paper saves about 17 trees? That's a win-win! Plus, it reduces the amount of waste going to landfills, which helps lower those pesky greenhouse gas emissions.

Let's not forget about your voice. Get involved in your community! Attend local meetings, support environmental initiatives, and advocate for policies that promote sustainability. Write letters to your representatives, join environmental groups, and participate in peaceful protests. It's amazing how one voice can inspire others to join the cause.

And hey, if you're feeling adventurous, consider going meatless one day a week. It's a little change that can make a BIG impact. Reducing your meat consumption can significantly lower your carbon footprint. Plus, your taste buds might thank you for trying out new plant-based recipes. You might discover that quinoa salad is your new jam!

As we wrap up this section, I want you to remember that change doesn't happen overnight. It's a journey, not a sprint. But every step you take counts. So, let's not let the weight of climate change crush our spirits. Instead, let's rise up, take action, and inspire others to do the same.

In conclusion, understanding climate change is crucial for all of us. The science is clear, and our human activities are contributing to the problem. But here's the kicker: YOU have the power to make a difference. It's time to roll up those sleeves and get to work. The planet needs us, and together, we can create a sustainable and bright future. So, what are you waiting for? Let's get started!

Now, let's dig deeper into how we can tackle this issue, one step at a time.

Understanding the Complexity of Climate Change

Climate change is like that complicated puzzle you just can't seem to solve. Each piece represents a different factor—industrialization, urbanization, deforestation, and more. To grasp the full picture, we need to understand how these pieces fit together and how they contribute to the bigger issue.

First, let's consider industrialization. The rise of factories and mass production has undoubtedly improved our lives in many ways, providing us with goods and services we couldn't have imagined before, but it has also led to a significant increase in greenhouse gas emissions. Factories burn fossil fuels to power their operations, releasing large amounts of CO_2 into the atmosphere. According to the Global Carbon Project, industrial processes account for about 21% of global emissions. That's a hefty chunk!

Now, think about urbanization. As more people flock to cities, we build more infrastructure—roads, buildings, and transportation systems. This growth often comes at the expense of green spaces and natural habitats. Trees and plants are removed to make way for concrete and asphalt, reducing the planet's ability to absorb CO_2. The result? More emissions, less biodiversity, and a whole lot of concrete.

Deforestation is another major player in this game. Forests act as carbon sinks, absorbing CO_2 and providing oxygen. They play a vital role in regulating the Earth's climate. When we cut down trees for timber or to clear land for agriculture, we're not just losing trees; we're losing a vital ally in the fight against climate change. According to the World Resources Institute, deforestation accounts for around 10% of global greenhouse gas emissions.

So, what can we do about it? Well, let's start with education. Knowledge is power, right? Understanding the impacts of our choices is the first step toward making better decisions. Whether it's choosing sustainable products, supporting eco-friendly companies, or advocating for policies that protect our environment, every action counts. Learning about the lifecycle of products, the environmental impact of different industries, and the policies that promote sustainability empowers us to make informed choices.

Next, let's talk about renewable energy. Transitioning from fossil fuels to renewable sources like solar, wind, and hydroelectric power is crucial for reducing our carbon footprint. Solar panels convert sunlight directly into electricity, wind turbines harness the power of the wind, and hydroelectric dams utilize the energy of moving water. According to the International Renewable Energy Agency (IRENA), renewable energy could supply up to 86% of global power needs by 2050. Imagine that! A world powered by clean, sustainable energy.

But here's the kicker: we can't do it alone. Governments, businesses, and individuals all play a role in this transition. Support policies that promote renewable energy, invest in green technologies, and encourage your friends and family to make sustainable choices. Advocate for incentives for renewable energy adoption, support companies that are investing in sustainable practices, and talk to your friends and family about the importance of switching to renewable energy sources.

Now, let's shift gears and talk about the role of technology in combating climate change. From electric vehicles to smart home devices, technology has the potential to revolutionize how we consume energy and reduce waste. Electric vehicles eliminate tailpipe emissions, contributing to cleaner air in our cities. Smart home devices, like programmable thermostats and smart lighting systems, help us conserve energy and reduce our carbon footprint. For example, did you know that smart thermostats can save you up to 15% on heating and cooling costs? That's not just good for your wallet; it's good for the planet too!

But let's not forget about the power of innovation. Entrepreneurs and inventors are stepping up to create solutions that tackle climate change head-on. From biodegradable packaging to carbon capture technologies, the possibilities are endless. Scientists are developing new ways to capture CO_2 from the atmosphere and store it underground, preventing it from contributing to global warming. Companies are creating biodegradable packaging made from plant-based materials, reducing the amount of plastic waste that ends up in landfills. So, keep an eye out for the next big idea that could change the game!

As we wrap up this section, remember that understanding climate change is just the beginning. It's a complex issue that requires a multifaceted approach. But by educating ourselves, supporting renewable energy, embracing technology, and encouraging innovation, we can make a real difference.

Taking Action: Your Role in the Fight Against Climate Change

Now that we've explored the science and complexity of climate change, it's time to get to the heart of the matter: what can YOU do about it? The truth is, every little action adds up, and together, we can create a ripple effect that leads to significant change.

Let's start with your daily habits. Consider how you can reduce your carbon footprint in your everyday life. Here are a few practical steps to get you started:

1. **Ditch Single-Use Plastics**: Bring reusable bags, water bottles, and containers wherever you go. Instead of buying bottled water, carry a reusable water bottle and fill it up at home or at water fountains. Instead of using plastic grocery bags, bring your own reusable bags to the store. Not only will you reduce waste, but you'll also save money in the long run.

2. **Opt for Public Transportation**: Whenever possible, take the bus, train, or bike instead of driving. Public transportation reduces the number of cars on the road, which in turn reduces greenhouse gas emissions. Biking and walking are not only good for the environment but also good for your health. You'll save on gas, reduce emissions, and get some exercise—all at the same time!

3. **Shop Local**: Support local farmers and businesses by purchasing food and products from your community. Local farmers often use more sustainable farming practices, and buying locally reduces the distance your food has to travel, which reduces transportation emissions. This reduces transportation emissions and helps your local economy thrive.

4. **Conserve Water**: Simple actions like taking shorter showers, fixing leaks, and using water-efficient

appliances can make a big difference. Every drop of water we save reduces the energy needed to treat and transport water. Remember, every drop counts!

5. **Educate Others**: Share what you've learned about climate change with friends and family. Start conversations, share articles, and encourage others to take action. The more people who are aware of the issue, the more likely we are to see real change.

6. **Advocate for Change**: Write to your local representatives, attend town hall meetings, and support policies that promote sustainability. Let your voice be heard and demand action on climate change. Your voice matters!

7. **Join or Start a Green Initiative**: Whether it's a community garden, a clean-up event, or a recycling program, getting involved in local initiatives is a great way to make a difference. Volunteering your time and energy can have a significant impact on your community.

8. **Reduce Energy Consumption**: Turn off lights when you leave a room, unplug devices you're not using, and consider switching to renewable energy sources for your home. These simple actions can significantly reduce your energy consumption and lower your carbon footprint.

9. **Embrace a Plant-Based Diet**: Even if you can't go fully vegetarian or vegan, consider incorporating more plant-based meals into your diet. Reducing your meat consumption can have a significant impact on the environment. Your taste buds—and the planet—will thank you!

10. **Stay Informed**: Keep up with the latest news on climate change and environmental issues. Knowledge is power, and being informed allows you to make better choices. Follow reputable sources of information and stay up-to-date on the latest developments.

As we wrap up this section, I want you to remember that change doesn't happen overnight. It's a journey, not a sprint. But every step you take counts. So, let's not let the weight of climate change crush our spirits. Instead, let's rise up, take action, and inspire others to do the same.

In conclusion, understanding climate change is crucial for all of us. The science is clear, and our human activities are contributing to the problem. But here's the kicker: YOU have the power to make a difference. It's time to roll up those sleeves and get to work. The planet needs us, and together, we can create a sustainable and bright future. So, what are you waiting for? Let's get started!

Together, we can tackle climate change, one step at a time.

Chapter 3

Reducing Waste

Plastic pollution? It's like that annoying guest who just won't take the hint and leave your party. It's everywhere, making a mess, and you can't ignore it. You've heard the numbers, right? Over 300 million tons of plastic get churned out each year. Imagine that—enough plastic to wrap around the Earth more than 1,500 times! The sheer volume is staggering, a testament to our reliance on this persistent material. But here's the kicker: only about 9% of that plastic gets recycled. The rest? It's chilling in landfills, taking up valuable space and leaching harmful chemicals into the ground; floating in oceans, forming massive garbage patches that disrupt marine ecosystems; and even making its way into the stomachs of sea turtles, who mistake plastic for food. Yikes! So, let's roll up our sleeves and dive into this issue headfirst.

First off, let's chat about how plastic pollution messes with our environment. It's not just about the aesthetics—though, let's be honest, a beach littered with plastic bottles isn't exactly a postcard scene. Imagine the once pristine sand now covered in plastic debris, the waves carrying plastic fragments instead of seashells. It's about the real damage that plastic does to our ecosystems. When plastic breaks down, it doesn't just vanish; it turns into microplastics. These tiny particles are everywhere—our soil, contaminating the very ground we walk on; our water, polluting our rivers, lakes, and oceans; and yes, even in the fish we chow down on. Studies show these microplastics can harm marine life, causing internal injuries and disrupting their digestive systems. Scientists are also studying the potential effects these microplastics could have on humans. Talk about a dinner party gone wrong!

So, what's the game plan? It starts with cutting back on single-use plastics in our daily grind. These are the plastics we use once and then throw away, like plastic bags, straws, and water bottles. Here are some strategies to help you kick that habit:

Ditch the plastic straws. Seriously, how many times have you grabbed a drink and thought, "I need a straw"? Probably not often. Instead, consider using a reusable straw made of stainless steel or silicone. They're easy to clean with a small brush and can be a conversation starter—"Look at my fancy straw!" Plus, they come in a variety of fun colors and designs.

Bring your own bags. Next time you hit the grocery store, toss a couple of reusable bags in your car. Keep them in your trunk or by the front door so you don't forget them. It's a small change, but it makes a big difference. Plus, you'll feel like a total eco-warrior as you stroll through the aisles, knowing you're doing your part to reduce plastic bag waste.

Opt for bulk. Buy in bulk when you can. Not only does it save you money, as you're often paying less per unit, but it also cuts down on packaging waste. Think about it: instead of buying ten individual bags of chips, grab one big bag. You can then portion them out into reusable containers at home. Your friends will thank you for the snacks, and your trash can will breathe a sigh of relief.

Say no to bottled water. Invest in a good reusable water bottle. Choose one made of stainless steel, glass, or BPA-free plastic. Not only will you save money, as you won't have to constantly buy bottled water, but you'll also keep those pesky plastic bottles out of the ocean. Plus, it's a great way to stay hydrated and look cool at the same time.

Get creative with food storage. Instead of using plastic wrap, try beeswax wraps or glass containers. Beeswax wraps are made of cotton coated in beeswax, and they can be reused multiple times. Glass containers are durable, easy to clean, and don't leach chemicals into your food. They're eco-friendly and keep your leftovers fresh. And hey, they look way better on your fridge!

Now that we've tackled some strategies, let's get a little creative. Upcycling and repurposing items isn't just a trend; it's a way to breathe new life into things that would otherwise end up in the trash. It's like giving your old stuff a second chance at life. Here are some fun ideas to get your creative juices flowing:

Mason jars galore. Those empty mason jars in your pantry? They're gold mines! Use them to store dry goods like rice, pasta, or beans; as vases for flowers, adding a touch of nature to your home; or even as candle holders, creating a warm and cozy atmosphere. You can even turn them into cute gifts by filling them with homemade goodies like jams, pickles, or baked treats.

T-shirt tote bags. Got some old t-shirts lying around? Don't toss them! Cut off the sleeves and neckline, turn them inside out, and sew the bottom shut. Boom! You've got yourself a stylish tote bag. Perfect for trips to the farmer's market or beach days. You can even decorate them with fabric paint or patches to personalize them.

Cork bulletin board. You can create a fun bulletin board by gluing corks to a piece of cardboard or a wooden board. You can use corks from craft stores or other sources. It's a great way to keep track of your to-do list and display photos.

Egg carton seed starters. Spring is just around the corner, and if you're planning to start a garden, don't toss those egg cartons! Fill them with soil, plant your seeds, and watch them grow. Once the seedlings are big enough, you can plant them directly in the ground, as the cardboard will decompose. It's a fantastic way to start your gardening journey while keeping waste to a minimum.

Furniture flip. Got an old piece of furniture that's seen better days? Instead of throwing it out, consider giving it a fresh coat of paint or new hardware. You'd be surprised how a little creativity can turn something drab into fab! You can find inspiration and tutorials online for transforming old furniture into stylish and functional pieces.

Now, I know what you're thinking: "But isn't it easier to just throw things away?" Well, yeah, it might be. It's the quick and easy option. But think about the long-term effects of that convenience. Every time you toss something in the trash, you're contributing to a cycle that harms our planet. And if you want to be part of the solution, it's time to roll up your sleeves and get your hands a little dirty—figuratively speaking, of course.

To sum it all up, reducing waste is about being mindful of our choices and taking small steps every day. It's about adopting a more conscious approach to consumption and disposal. The impact of plastic pollution is serious, but we have the power to make a difference. By minimizing single-use plastics and getting creative with upcycling, we can all contribute to a healthier planet.

So, here's your challenge: pick one strategy from the list above and implement it this week. Whether it's ditching plastic straws or creating a tote bag from an old t-shirt, every little bit counts. And remember, it's not about being perfect; it's about making progress. Small changes, when adopted by many, can have a significant impact. Together, we can turn the tide on plastic pollution and create a brighter future for our planet. Let's do this!

Alright, let's dive deeper into this whole waste-reduction gig. It's not just about a few changes here and there; it's about creating a lifestyle shift. You know, it's like switching from a flip phone to a smartphone. At first, it feels like a hassle, learning all the new features and navigating the touchscreen, but once you get the hang of it, you wonder how you ever lived without it.

Think about your daily routine. You wake up, brush your teeth, and grab breakfast. How much waste do you create in those simple morning tasks? That's where you can start making changes. Swap out your plastic toothbrush for a bamboo one, which is biodegradable. Use a reusable coffee cup instead of those paper or plastic ones from the café. Many cafes even offer discounts for customers who bring their own cups. It's all about those little tweaks that stack up over time.

And let's not forget about the power of community. Getting your friends and family involved can make a huge difference. Organize a cleanup day at your local park or beach. You'll not only be doing something good for the environment, removing litter and debris, but you'll also have a blast, spending time with friends and making a positive impact. Plus, it's a great way to spread the word about reducing waste, inspiring others to join the cause. Who knows? Maybe you'll inspire someone to make changes in their own life.

Speaking of community, let's talk about local businesses. Supporting them can really help cut down on waste. Buy your fruits and veggies from a local farmer's market instead of the grocery store. Not only will you get fresher produce, often grown without harmful pesticides, but you'll also be reducing the carbon footprint associated with transporting food over long distances. It's a win-win!

Now, I know some folks might be thinking, "But isn't it more expensive to go green?" Well, it can be, but it doesn't have to be. Sure, organic foods and fancy reusable products can cost a pretty penny. But you can save money by buying seasonal produce, which is often cheaper and more readily available, use coupons, and look for bulk items. Buying in bulk often reduces the cost per unit. Plus, think about all the cash you'll save by not buying single-use plastics. That adds up!

And let's chat about meal planning. It's a game changer! Planning your meals for the week can help you avoid impulse buys and reduce food waste. You'll know exactly what you need, creating a grocery list and sticking to it, and you won't end up with a fridge full of rotting veggies. Not to mention, it saves you time during the week. Who doesn't love a little extra time?

Now, here's a fun thought: how about starting a compost bin? It's a fantastic way to reduce food waste and create nutrient-rich soil for your garden. You'd be surprised at how much food scraps can add up. Things like fruit and vegetable peels, coffee grounds, and eggshells can be composted. There are different ways to compost, even if you don't have a yard, so it's worth looking into!

Speaking of gardens, let's talk about growing your own food. It's easier than you think! Even if you don't have a backyard, you can grow herbs on your windowsill or in small pots. It's a great way to reduce packaging waste from store-bought herbs and spices. Plus, there's something incredibly satisfying about cooking with ingredients you've grown yourself. You can even start small with easy-to-grow plants like lettuce, tomatoes, or peppers.

And don't forget about the clothes you wear. The way many clothes are made today uses a lot of resources and can create a lot of waste when clothes are thrown away quickly. Instead of buying new clothes every season, try thrifting or swapping clothes with friends. It's like a treasure hunt! You never know what gems you might find. Plus, it's a fun way to refresh your wardrobe without contributing to the waste problem.

And here's a little secret: you don't have to do it all at once. Start small. Pick one or two changes to implement this week, and build from there. Maybe it's as simple as using a reusable bag for groceries or cutting back on takeout containers. Every small step counts, and over time, these small changes can become lasting habits. It's all about progress, not perfection.

Now, let's get real for a second. Change can be tough. It's easy to slip back into old habits, especially when life gets busy. But every little bit helps. If you have a slip-up, don't beat yourself up about it. Just dust yourself off and keep going. It's all part of the journey. The key is to be persistent and to keep the bigger picture in mind.

So, as we wrap this up, remember that reducing waste is a marathon, not a sprint. It takes time, effort, and a little creativity. But the payoff? It's huge. A cleaner planet, healthier ecosystems, and a brighter future for generations to come. By reducing waste, we're not only protecting the environment but also conserving valuable resources and creating a more sustainable future.

So, what's it going to be? Are you ready to take that first step? Pick a strategy, get your friends involved, and start making those changes.

The planet will thank you for it. Let's roll up our sleeves and get to work—together, we can make a real difference!

Chapter 4

Sustainable Eating

When it comes to saving the planet, what we chow down on can really shake things up. Seriously! Ever thought about how your lunch could be doing the Earth a solid? Let's dive into the tasty world of sustainable eating. Buckle up, 'cause we're about to explore the perks of a plant-based diet, figure out what food miles are all about, and share some nifty tricks for cutting down on food waste at home. Grab a snack—maybe something green—and let's roll!

First off, let's chat about why a plant-based diet is a game changer. You might be thinking, "Why should I trade my juicy cheeseburger for a salad?" Well, here's the lowdown: switching to a plant-based diet is like giving Mother Nature a high-five. Studies show that if everyone cut back on meat and dairy, we could slash greenhouse gas emissions big time. The United Nations even says that if we all hopped on the plant-based train, we could drop food-related emissions by up to 70% by 2050. That's no small potatoes—literally! Imagine the impact if everyone made even small changes.

Now, I get it. You might be picturing a life with no flavor or satisfaction. But let me tell you, there's a whole smorgasbord of deliciousness out there. Think of it this way: swapping out meat doesn't mean giving up taste; it's like trading in your old flip phone for a slick smartphone. You're just leveling up your food game! From hearty lentil stews to colorful veggie tacos, the options are endless. Picture vibrant colors, enticing aromas, and satisfying textures. Plus, munching on more plants can boost your health—lowering your risk of heart disease, diabetes, and even some cancers. So, not only are you saving the planet, but you're also doing your body a solid. Win-win, right? It's a change that benefits both you and the world around you.

Now, let's switch gears and tackle food miles. What the heck are food miles, you ask? Well, it's the distance your food travels from farm to table. Picture this: that avocado toast you love? If it's flown in from halfway around the globe, it's racked up some serious food miles. And guess what? Those miles contribute to carbon emissions. It's like your brunch is doing the cha-cha with climate change. Every mile traveled requires fuel, which releases harmful greenhouse gases into the atmosphere.

So, how can you cut down on those pesky food miles? Easy peasy! Start by shopping local. Hit up your farmer's market or join a community-supported agriculture (CSA) program. You'll score fresh, seasonal produce, and you'll be supporting local farmers. Plus, you'll strut around like a true eco-warrior with your reusable tote bag stuffed with local goodies. Talk about a confidence boost! Imagine the vibrant stalls, the fresh smells of produce, and the friendly faces of the farmers.

But wait, there's more! Getting the hang of seasonal eating is another way to trim those food miles. If you're munching on strawberries in December, those berries have traveled a long way to get to your plate. Instead, embrace the rhythm of the seasons. Eat what's in season, and you'll not only enjoy fresher flavors, but you'll also be making a positive impact on the environment. Think crisp apples in the fall, juicy tomatoes in the summer, and hearty root vegetables in the winter.

Now, let's tackle a biggie: food waste. Did you know that about one-third of all food produced globally is wasted? That's like tossing away a third of your paycheck every month. Crazy, right? Not only is that a bummer for your wallet, but it's also a huge problem for our planet. When food rots in landfills, it releases methane—a greenhouse gas that's even nastier than carbon dioxide. So, how do we curb this waste? Here are some practical tips to get you started:

First up, plan your meals. Before you hit the grocery store, whip up a meal plan. Write down what you need for the week and stick to it. Trust me, your future self will thank you when you're not staring at a fridge full of sad, wilting veggies. This simple step can save you money and reduce waste.

Next, get creative with leftovers. Don't let those leftovers go to waste! Turn last night's roasted veggies into a frittata or toss them into a soup. It's like a culinary game of Tetris—fit those pieces together for a delicious meal! There are countless recipes online for turning leftovers into exciting new dishes.

Also, know your expiration dates. "Best by," "sell by," and "use by" can be super confusing. Don't let those labels dictate what goes in the trash. Often, food is still perfectly good to eat after the date. Use your senses—if it smells fine and looks okay, it's probably still good! These dates are often about peak quality, not safety.

And hey, if you've got extra fruits or veggies, freeze them! You can toss them in smoothies or soups later. It's like hitting the pause button on food waste. Freezing preserves the food and allows you to use it later, preventing it from spoiling.

Lastly, if you can't eat it, compost it! Not only does composting reduce waste, but it also enriches the soil. You'll be giving back to Mother Earth while feeling like a gardening guru. Composting turns food scraps into nutrient-rich soil that can be used to grow more food.

As we wrap up this chat about sustainable eating, remember that every little bit counts. You don't have to go full vegan overnight—just take small steps. Swap one meat meal for a plant-based option each week. Start shopping local. Get creative with your leftovers. These small changes can make a big difference over time.

The journey to sustainable eating is like a road trip. You don't need to rush to the destination; enjoy the ride! And hey, if you slip up, don't sweat it. Just like my attempts at baking—sometimes it's a total flop, but that's part of the learning process. It's all about progress, not perfection.

So, what's your next move? Take a moment to reflect on your eating habits. What small change can you make today? Remember, you've got the power to make a difference, one bite at a time. Let's chow down on sustainability and make our planet a better place!

Now, let's dive a little deeper into some personal stories and hypothetical scenarios to really drive the point home.

Imagine you're at a barbecue with friends. The grill's fired up, and there's a mountain of burgers sizzling away. You're eyeing that juicy beef patty, but then you remember the stats about meat consumption and its impact on the environment. You take a deep breath and decide to try a black bean burger instead. Your friends raise their eyebrows, but you're feeling adventurous.

That first bite? Oh man! It's smoky, spicy, and packed with flavor. You're surprised by how good it is. By the end of the night, you're sharing the recipe with your friends, and they're raving about how they'd never thought a veggie burger could taste so good. Just like that, you've not only enjoyed a delicious meal but also sparked a conversation about sustainable eating. Who knew your choice could inspire others?

Or picture this: it's Sunday, and you're planning your meals for the week. You pull out a cookbook filled with plant-based recipes. You start to get excited about trying new dishes. You pick a couple of recipes that use seasonal ingredients from your local farmer's market.

When you hit the market, you're greeted by the friendly farmer who grows your favorite veggies. You load up your basket with fresh produce and feel good knowing you're supporting local agriculture. As you prepare your meals throughout the week, you notice how vibrant and tasty everything is. You even have some leftovers that you creatively repurpose into new meals.

Now, let's talk about those food miles again. You've been trying to eat more locally, and it's opened your eyes to the incredible variety of food available in your area. You start looking forward to visiting the farmer's market every week. You discover new fruits and veggies you've never tried before.

It's like a treasure hunt! You chat with vendors, learn about their farming practices, and feel a sense of community. You realize that eating sustainably isn't just about the food;

it's about connecting with the people who grow it. It's a whole vibe, and you're here for it.

And what about food waste? You're in the kitchen, and you've got a bunch of leftover veggies staring at you. Instead of letting them go bad, you decide to whip up a stir-fry. You throw in whatever you've got—some wilting spinach, a couple of carrots, and maybe a bell pepper that's been lurking in the fridge for a while. You toss in some soy sauce, garlic, and ginger, and before you know it, you've created a delicious meal that not only tastes great but also saved those veggies from the trash. You feel like a culinary genius!

Now, let's get real. We all have those days when we just don't feel like cooking. It's easy to grab takeout or hit up the drive-thru. But what if you made a pact with yourself? Just one night a week, you cook something from scratch. You'll feel accomplished, and you'll know exactly what's going into your food. Plus, you can invite friends over for a little dinner party. Imagine the satisfaction of creating a meal with your own hands, knowing you're making a healthy and sustainable choice.

Picture this: you've invited a few pals over, and you're whipping up a plant-based feast. You make a killer pasta primavera loaded with fresh veggies and a creamy cashew sauce. Everyone's digging in, and they can't believe it's all plant-based. You're the host with the most, and you've turned them onto the idea of sustainable eating without even trying. The aroma of fresh herbs and vegetables fills the air, and the table is alive with conversation and laughter. You've created a positive experience that showcases the deliciousness of sustainable food.

And hey, let's not forget about those moments when you mess up. You tried to make a fancy vegan dessert, and it turned out to be a disaster. Instead of feeling defeated, you laugh it off. You share the story with your friends, and they appreciate your honesty. It's a reminder that nobody's perfect, and we're all just trying to do our best. Maybe the cake fell flat, or the sauce was too runny, but the important thing is that you tried. These experiences are all part of the learning process.

So, what's your takeaway from all this? Sustainable eating doesn't have to be an all-or-nothing deal. It's about making conscious choices that align with your values. It's about finding joy in cooking, discovering new flavors, and connecting with your community. It's about understanding the impact of your food choices and making informed decisions.

As you think about your next meal, consider how you can make it a little more sustainable. Maybe it's choosing a plant-based option, shopping at the local market, or getting creative with leftovers. Whatever it is, know that you're part of a bigger movement. Every little change adds up, and together, we can make a real impact. Think of it as a ripple effect – your small actions can inspire others to make similar changes.

So, let's chow down on sustainability! The planet will thank you, and who knows—you might just discover a new favorite dish along the way.

You'll not only be contributing to a healthier planet but also enriching your own life with new culinary experiences and a deeper connection to your food.

Chapter 5

Energy Efficiency

Alright, let's get real for a sec. Energy use is kind of like that friend who just won't leave the party. You know they're causing a ruckus, but kicking them out? Yeah, that's tough. The reality is, energy consumption is a huge factor in climate change. It's behind a staggering 70% of greenhouse gas emissions in the U.S. alone. That's no small potatoes, folks. And every time you flick a switch or crank the AC, you're adding to that mess. But don't freak out! There are plenty of simple steps you can take to chill out climate change right from your living room. Think of it this way: every small action you take at home, when multiplied by millions of homes, can make a real difference in reducing our collective impact on the planet.

So, let's dig into some easy tweaks you can make to ramp up your home's energy efficiency. This isn't rocket science, but it might feel like it when you realize just how simple it can be. Here's a quick rundown to kick things off:

First up, seal those drafts. Seriously, check around your windows and doors. Feel for a cool breeze seeping in even when they're closed tight? That's a draft. A little caulk or weather stripping can save you some serious cash on your energy bill. Think of it like wrapping your house in a cozy blanket—keeps the warmth in during winter and the cool air in during summer. This simple act prevents your heating and cooling systems from working overtime, saving energy and money, and making your home more comfortable.

Next, let's talk lighting. Ditch those old-school incandescent bulbs for LED ones. They're like the superheroes of light bulbs—using up to 80% less energy and lasting way longer. Plus, they come in all sorts of funky colors. Who wouldn't want a purple glow in their living room? Imagine the difference in your energy bill over time—and fewer trips to the store to buy replacements. LEDs also produce less heat, which further contributes to energy savings.

Now, here's a fun one: unplug your electronics when you're not using them. Those chargers and devices are sneaky little energy suckers, even when they're off. It's like putting your phone on airplane mode—less energy, more chill. These "phantom loads," also known as vampire power, might seem insignificant individually, but they add up over time, draining energy and costing you money.

Ever heard of a programmable thermostat? If not, you're missing out! Set it to lower the temp when you're out. You'll save energy and cash, and it's like having a personal assistant for your heating and cooling. Imagine setting it to automatically adjust the temperature throughout the day—warmer when you're home and active, cooler when you're at school or asleep. This smart technology can significantly reduce energy waste.

And let's not forget about insulation. If your house is older than your grandma's favorite sweater, it might need some love. Good insulation keeps your home comfy year-round. It's like giving your house a warm hug. Proper insulation in the walls, attic, and floors creates a thermal barrier, preventing heat from escaping in the winter and entering in the summer, thus reducing the need for excessive heating and cooling.

Now, let's chat about the perks of renewable energy sources. You've probably heard of solar panels and wind turbines, but what's the deal with them? Well, they're the rockstars of the energy scene. Here's why they rock:

First off, they're sustainable. Renewable energy comes from sources that won't run out—sunshine, wind, and water. It's like having an endless buffet of energy. Who doesn't love a buffet? Unlike fossil fuels, which are finite and contribute to pollution, these sources are constantly replenished by natural processes.

Then there's the money side. Once you invest in renewable energy, like solar panels, you can seriously slash your energy bills. It's like finding cash in your old jeans—what a surprise! Over the long term, the savings can be substantial, often offsetting the initial investment and even generating income through net metering programs.

And let's not forget job creation. The renewable energy sector is booming. Investing in it means you're not just saving the planet; you're helping create jobs. It's a total win-win! From manufacturing and installation to research and development, the renewable energy industry offers diverse and growing employment opportunities.

Also, relying on renewables means less dependence on fossil fuels. It's like cutting ties with that toxic ex—you'll feel lighter and free. Reducing our reliance on fossil fuels decreases air and water pollution, lessens our vulnerability to price fluctuations, and promotes energy independence.

Now, I know what you're thinking. "This all sounds great, but how do I actually make these changes?" Let's break it down into bite-sized steps.

Start small. Pick one or two energy-efficient changes to tackle this week. Maybe switch to LED bulbs or seal those pesky drafts. Once you've got those down, move on to the next. It's like leveling up in a video game—each small win gets you closer to the ultimate boss fight: a sustainable home!

Next, do a little digging into renewable energy options in your area. Check out local incentives for solar panel installation or community wind projects. You might be surprised at what's available. Many local governments, utility companies, and even federal programs offer rebates, tax credits, and other incentives to encourage the adoption of renewable energy.

And hey, don't hesitate to ask for help! Chat with friends or family who've made similar changes. You can learn a ton from their experiences, and they might even swing by to help you with the installation—free labor, anyone? Sharing knowledge and collaborating with others can make the transition to energy efficiency easier, more affordable, and more fun.

Here's a fun fact for you: if every American home switched to energy-efficient lighting, we could save about $6 billion a year, according to the U.S. Department of Energy. That's a whole lot of pizza money! Just think of what you could do with that cash—maybe hit the beach or treat yourself to that fancy coffee you've been eyeing. That money could also be reinvested in other sustainable initiatives or used to support local businesses.

But let's not forget the emotional side of this journey. Making your home more energy-efficient isn't just about saving bucks or cutting your carbon footprint; it's about creating a sense of pride. When you take action, you're not just sitting on the sidelines—you're in the game of sustainability. You're taking control, and that feels pretty darn good. It's empowering to know that you're actively contributing to a healthier planet and a more sustainable future.

In the end, energy efficiency isn't just some buzzword; it's a lifestyle. It's about making choices that benefit not only you but also the planet. And while it might feel overwhelming at times, remember that every little bit counts. You don't have to change the world overnight. Just take it one step at a time, and before you know it, you'll be cruising toward a more sustainable future.

So, what are you waiting for? Get out there and start making those changes! Your future self will thank you, and so will the planet. Let's make energy efficiency the norm, not the exception. Together, we can create a brighter, greener future for generations to come.

Now, let's dig a bit deeper into some of these ideas. Energy efficiency isn't just about making changes in your own home. It's also about spreading the word and getting your community involved. Think about it: if everyone pitches in, the impact could be massive. Imagine if your whole neighborhood got on board with energy-saving measures. You'd have a community of eco-warriors, all working together to lower energy consumption and fight climate change.

You could even start a local initiative—maybe a neighborhood challenge to see who can reduce their energy use the most. A little friendly competition never hurt anyone, right? Plus, it's a great way to bond with your neighbors. You could share tips, swap stories, and even host workshops on energy efficiency. Who knows, you might even inspire someone to make bigger changes, like installing solar panels or starting a community garden.

Speaking of community gardens, let's talk about the connection between energy efficiency and sustainable living. It's all intertwined. When you start being mindful about energy use, you often find yourself making other eco-friendly choices too. Maybe you'll start composting, using reusable bags, or even growing your own veggies. It's like a snowball effect—once you start, it just keeps rolling. Each positive change reinforces the others, creating a more holistic and sustainable lifestyle.

And let's be real, there's something really satisfying about living sustainably. It's like you're part of a bigger movement, contributing to something greater than yourself. You're not just another face in the crowd; you're making a difference. Plus, you'll feel a sense of accomplishment every time you see your energy bill drop or notice how comfy your home is thanks to those insulation upgrades. It's a rewarding feeling to know that your actions are having a tangible positive impact on the environment and your own life.

Now, I get it—some folks might feel a bit lost when it comes to these changes. Maybe you're thinking, "I don't have the time or money for this." But hear me out. You don't have to overhaul your entire life in one go. Start with what you can manage. Maybe it's just changing a few light bulbs or adjusting your thermostat. Every little bit helps. Even small, affordable changes can add up to significant savings over time.

And let's not forget about the resources out there. There are tons of online guides, videos, and community programs that can help you navigate this journey. Don't hesitate to tap into those. YouTube is a goldmine for DIY energy-saving tips, and local energy companies often offer incentives for making your home more efficient. Many organizations and government agencies also provide free or low-cost energy audits to help you identify areas for improvement in your home.

Oh, and let's not overlook the tech side of things. Smart home devices can make energy efficiency a breeze. From smart thermostats to energy monitors, there are gadgets out there that help you track and manage your energy use. It's like having a personal trainer for your energy consumption—keeping you on track and motivated. These devices can provide valuable insights into your energy usage patterns, allowing you to make informed decisions and optimize your energy consumption.

Now, let's switch gears for a second and talk about the future. What does it look like if we all commit to energy efficiency? Imagine a world where clean energy is the norm. Picture cities powered by solar and wind, where homes are designed to be energy-efficient from the ground up. It's a world where we're not just surviving but thriving, living in harmony with our environment. Imagine rooftop gardens on every building, electric vehicles silently gliding through the streets, and the air filled with the sounds of nature, not traffic congestion.

And here's the kicker: it's not just a dream. It's entirely possible if we all pitch in. Governments, businesses, and individuals can work together to make this vision a reality. It'll take time, effort, and a lot of collaboration, but it's doable. Think of the positive impact on future generations – a cleaner, healthier planet for them to inherit, with thriving ecosystems and a stable climate.

So, let's rally together! Let's make energy efficiency a priority in our lives and communities. Talk about it with your friends, share what you've learned, and encourage others to join the cause. The more people who get involved, the bigger the impact we'll have. Start a conversation at school, with your family, or in your social circles. You might be surprised at how many people are interested in making a difference and contributing to a better future.

And remember, it's not about being perfect. It's about making progress. Every step you take, no matter how small, counts. So don't get discouraged if you can't do it all at once. Just keep moving forward, and celebrate those victories along the way. Even small changes, like remembering to turn off lights when you leave a room or using a power strip to easily switch off multiple devices, contribute to the bigger picture of a more sustainable world.

In conclusion, energy efficiency is more than just a buzzword—it's a way of life. It's about making smart choices that benefit both you and the planet.

From sealing drafts to investing in renewable energy, every action counts. So, roll up your sleeves, get involved, and let's make a difference together. Your future self—and the planet—will be grateful.

Chapter 6

Eco-Friendly Transportation

Alright, folks, let's hit the road—figuratively speaking, of course. We're diving into the world of eco-friendly transportation, where every mile counts, and every choice matters. So buckle up! Or, better yet, let's ditch the car keys for a moment and explore some greener options. Let's consider how our daily commutes and travels impact the planet and what we can do to lessen that impact.

First off, let's talk about the environmental impact of different modes of transport. It's no secret that cars are like that friend who shows up uninvited and eats all your snacks. They guzzle gas, pump out emissions, and contribute to air pollution faster than you can say "carbon footprint." According to the Environmental Protection Agency, transportation accounts for nearly 29% of greenhouse gas emissions in the U.S. That's a hefty slice of the pie! These emissions contribute to climate change, impacting air quality and overall environmental health.

Now, let's not forget about the alternatives. Biking, walking, and public transit are like the superhero trio of sustainable travel. Each one comes with its own set of benefits that not only help the planet but can also make your life a whole lot easier. These options offer practical and positive ways to reduce your environmental impact while also improving your well-being.

Biking, for instance, is not just a great way to tone those legs and feel the wind in your hair. It's a low-impact exercise that gets your heart pumping and your mood soaring. Plus, it's cheaper than gas! You ever try to fill up a bike? Yeah, didn't think so. You can even find some rad bike-sharing programs in many cities that let you rent a bike for a quick trip. Talk about a win-win! Imagine cruising down the street on a bike, feeling the fresh air and getting some exercise, all while reducing pollution.

Walking? Oh man, it's like the original form of transportation. It's free, it's simple, and it's a fantastic way to clear your head. Think about it: when you walk, you're not just getting from point A to point B. You're soaking in the sights, sounds, and smells of your neighborhood. You're connecting with your community. And hey, it's a great way to sneak in some exercise without even trying! From the chirping of birds to the bustling of local shops, walking allows you to experience your surroundings in a much more intimate way.

Now, public transit is like the underdog of transportation. It's often overlooked, but it's got some serious potential. Taking the bus or train can significantly reduce your carbon emissions, especially if you're traveling with a group. Plus, you can kick back, relax, and scroll through your phone instead of stressing over traffic. Just think about how many hours you could save each week if you didn't have to deal with rush hour. Imagine using that time to read a book, listen to music, or catch up with friends.

But hold on a second! What if your community doesn't have great transportation options? Well, that's where YOU come in. Advocacy is the name of the game, my friend. You've got the power to push for better transportation options in your area. Start by getting involved in local meetings or joining community groups focused on sustainability. Bring your friends along—strength in numbers, right? By working together, you can amplify your voice and make a real difference.

Here's a quick checklist of steps you can take to advocate for better transportation:

1. **Research**: Find out what transportation options are currently available in your community. What's working? What's not? Look into bus routes, train schedules, bike lanes, and pedestrian walkways. Identify areas where improvements are needed.

2. **Connect**: Reach out to local leaders, transit authorities, or even neighborhood associations. Express your concerns and ideas. Contact your city council members, transportation planners, and community organizations to voice your support for better transportation.

3. **Gather Support**: Talk to your friends, family, and neighbors. The more voices you have, the louder your message will be. Organize petitions, write letters to local newspapers, and spread the word on social media to build a strong base of support.

4. **Attend Meetings**: Get involved in city council meetings or public forums. Don't be shy! Your input matters. Participating in these meetings allows you to directly address decision-makers and contribute to the conversation.

5. **Use Social Media**: Create a buzz online. Share your thoughts, ideas, and progress. Hashtags can be your best friend here. Use platforms like Twitter, Facebook, and Instagram to raise awareness, connect with other advocates, and share information about local transportation initiatives.

6. **Stay Persistent**: Change takes time. Keep pushing for what you believe in. Every small step counts! Don't get discouraged if you don't see results immediately. Consistent effort and dedication are key to achieving long-term change.

Now, let's take a moment to reflect on how these changes can impact not just the environment but also your life. Imagine a world where biking or walking is the norm. Picture cleaner air, less traffic, and communities that thrive on connection. Sounds dreamy, right? But it's not just a fantasy. It's within our reach if we take action. Imagine quieter streets, more green spaces, and a stronger sense of community as people interact with each other more frequently.

You might be thinking, "But what if I can't bike or walk everywhere?" That's totally fair! Not everyone can ditch their car completely, and that's okay. The goal here is to make small, manageable changes that fit your lifestyle. Even swapping out one car trip a week for public transit or a bike ride can make a difference. Consider combining different modes of transport for longer journeys or using your car only when absolutely necessary.

Let's break it down even further. Here are some practical tips for incorporating eco-friendly transportation into your routine:

Plan Your Routes: Use apps like Google Maps or Waze to find the most efficient routes, whether you're walking, biking, or using public transit. These apps can also provide information on bus and train schedules, bike paths, and walking routes.

Buddy Up: Carpool with friends or neighbors. Not only does it save on gas, but it's also a great way to catch up. Sharing rides can also reduce traffic congestion and parking issues.

Invest in Gear: If you're biking, invest in a good helmet and lights. Safety first, folks! Proper safety gear is essential for protecting yourself while biking, especially at night or in low-light conditions.

Mix It Up: Try combining different modes of transport. Walk to the bus stop, take the bus to a train station, and then bike the rest of the way. This can be a great way to cover longer distances while still incorporating eco-friendly options.

Stay Informed: Keep an eye on local transportation initiatives. New bike lanes or improved public transit options might be on the horizon! Stay updated on local news and transportation plans to take advantage of new opportunities for sustainable travel.

Now, I know what you're thinking: "But what if my friends and family don't get on board?" That's a tough nut to crack, but remember, change often starts with one person. Share your experiences and the benefits you've seen. Sometimes, all it takes is a little inspiration to get others excited about making a change. Lead by example and show them how easy and rewarding it can be to adopt eco-friendly transportation habits.

So, what's the takeaway here? Eco-friendly transportation isn't just about saving the planet; it's about enhancing your quality of life. It's about connecting with your community, staying active, and making conscious choices. It's about creating healthier, more vibrant communities for everyone.

And let's be real—life is too short to sit in traffic. So, let's roll up our sleeves, hop on our bikes, and hit the ground running. Every step, every pedal, and every bus ride counts. Let's make our communities greener, one mile at a time. Let's create a future where sustainable transportation is the norm, not the exception.

Whether it's biking, walking, or using public transit, you've got the power to make a difference.

And remember, every journey starts with a single step—or pedal! So what are you waiting for? Let's get moving!

Chapter 7

Water Conservation

You ever turn on the faucet and think, "Man, this water's just flowing like my hopes and dreams?" Well, it's time to hit the brakes on that water wastage. Conserving water isn't just a good idea; it's a necessity. We're talking about a resource that's as vital as your morning coffee (or whatever gets you going!). Without it, we're in a world of hurt. So, let's dive into why saving water is important, how you can do it at home, and the bigger picture of the global water crisis. Consider the sheer volume of water used every day – from brushing our teeth and taking showers to watering lawns and manufacturing the products we use – and the cumulative impact of even small reductions in usage.

First off, let's chat about why conserving water should be a part of your daily life. Think of water as the lifeblood of our planet. It's not just for drinking and showering—it's in our food (think about how much water it takes to grow fruits and vegetables or raise livestock), our clothes (cotton farming is a huge water consumer), and even in the tech we use (manufacturing electronics requires vast amounts of water). When you waste water, you're not just wasting a resource; you're wasting energy (it takes energy to treat and transport water), money, and time. It's like pouring your paycheck down the drain—literally! And if that doesn't motivate you, consider this: over 2 billion people worldwide lack access to safe drinking water. That's a staggering number, and it's rising. Imagine the daily struggles of those who have to walk miles for clean water, often putting themselves at risk, or who lack basic sanitation, leading to serious health problems.

So, what can you do about it? Here's a handy list of practical tips to reduce water usage at home:

1. **Shorten Your Showers**: Aim for five minutes. Set a timer if you have to! You'd be surprised how much water you can save. Plus, it's a great excuse to skip the whole "I'll just stand here and think about life" routine. Those long, steamy showers might feel relaxing, but they use a significant amount of water – think about filling up a bathtub multiple times for just one long shower.

2. **Fix Leaks**: A dripping faucet might not seem like a big deal, but that little drip can waste over 3,000 gallons a year. That's enough water for a year's worth of drinking for a family of four. Get those leaks fixed, stat! A simple leaky faucet can waste enough water to fill a small swimming pool over a year. Even a running toilet can waste hundreds of gallons a day!

3. **Turn Off the Tap**: When brushing your teeth, washing your hands, or doing dishes, turn off the tap. It's simple

but effective. Just think of it as a mini water-saving challenge! Every time you leave the water running unnecessarily, you're contributing to water waste. It's a small change that makes a big difference.

4. **Use a Broom, Not a Hose**: When cleaning driveways or sidewalks, use a broom instead of hosing it down. It's a workout and saves a ton of water. Two birds, one stone! Hosing down surfaces uses a surprising amount of water, and often, a broom is just as effective, if not more so, at removing dirt and debris.

5. **Water-Efficient Appliances**: If you're in the market for new appliances, look for those that are water-efficient. They may cost a bit more upfront, but they'll save you money in the long run. Plus, they're better for the planet—win-win! Look for the WaterSense label, which indicates that an appliance meets EPA criteria for water efficiency. This includes washing machines, dishwashers, and even toilets.

6. **Collect Rainwater**: If you have a garden, consider installing a rain barrel. You can use that water for your plants instead of the tap. Nature's free gift to you! Rainwater is a great natural source of water for your garden and doesn't contain chemicals often found in tap water. You can even use it to wash your car!

7. **Mind Your Lawn**: Water your lawn in the early morning or late evening to minimize evaporation. And consider

xeriscaping—using drought-resistant plants that require less water. Your yard will look great, and you'll save water. Watering during cooler parts of the day reduces water loss due to evaporation from the sun and heat. Native plants are often a great choice for xeriscaping as they are adapted to the local climate and require minimal watering.

8. **Educate Others**: Share your water-saving tips with friends and family. You'd be surprised how much of an impact you can have just by spreading the word. Talking about water conservation can raise awareness and encourage others to make changes in their own lives. Start a conversation at school or with your social group.

Now, let's take a step back and look at the global water crisis. This isn't just a problem for far-off places; it's knocking on our doors too. Climate change, population growth, and pollution are all playing a part in this mess. According to the United Nations, by 2025, 1.8 billion people will live in areas with absolute water scarcity. Think about that for a second. That's like trying to find a good parking spot at a packed concert—nearly impossible! Imagine the impact this has on agriculture, sanitation, and overall quality of life. This can lead to food shortages, disease outbreaks, and even social unrest.

The implications are serious. Water scarcity can lead to conflict, economic downturns, and health crises. And if we don't change our ways, it's only going to get worse. This is why your individual actions matter. Every drop saved counts. You may think, "I'm just one person; what can I do?" Well, you can do a lot! Even small changes in your daily habits can contribute to a larger positive impact. Think of it as a collective effort – every little bit helps.

Let's not forget about the environment. When we conserve water, we're also protecting ecosystems. Rivers, lakes, and wetlands all rely on a delicate balance of water. Overuse can lead to habitat destruction, which affects wildlife. So, when you save water, you're not just helping yourself; you're helping the planet. It's like being a superhero, but instead of a cape, you've got a reusable water bottle. Protecting these ecosystems is essential for maintaining biodiversity and the overall health of the planet. These ecosystems provide vital services, such as filtering water and providing habitats for countless species.

So, here's a challenge for you: pick one of those tips and implement it this week. Maybe it's fixing that leaky faucet you've been ignoring or setting a timer for your showers. Whatever it is, make it a habit. Track your progress and see how much water you can save in a month. Share your results with friends and inspire them to join in. Turning these actions into regular habits is key to making a long-term difference. You could even create a competition with your friends to see who can save the most water!

In conclusion, conserving water is essential for our daily lives and the future of our planet. It's not just about saving money or being eco-friendly; it's about ensuring that everyone has access to this precious resource.

So, let's roll up our sleeves and get to work. Together, we can make a difference—one drop at a time. And remember, every small action adds up to big changes.

Now go forth and be the water-saving warrior you were born to be!

Chapter 8

Supporting Sustainable Brands

Alright, everyone, let's jump right into the world of sustainable brands. You might be scratching your head, wondering, "Why should I care?" Well, let me tell you, it's a big deal! Supporting eco-friendly products and companies isn't just some passing fad; it's a lifestyle change that can help us save our precious planet. So, grab a snack—maybe something plant-based—and let's dig in! Consider how the products we buy and the companies we support influence the health of our planet and the well-being of communities around the world.

First off, how do you even find eco-friendly products and companies? It can feel like hunting for a needle in a haystack sometimes, but don't sweat it! Here are some straightforward tips to help you sift through the chaos:

Look for Certifications: When you spot labels like USDA Organic, Fair Trade, or Energy Star, you know you're onto something solid. These certifications are like gold stars for companies doing their part for Mother Earth. It's like discovering your favorite band is actually super eco-conscious. Who knew? These labels provide a quick and easy way to identify products that meet specific environmental or social standards.

Research the Brand's Values: Most companies have an "About Us" page on their website. Go ahead, check it out! If they're boasting about their sustainable practices, that's a good sign. But if they're just rambling about how long they've been around without mentioning their environmental impact, maybe it's time to keep looking. Look for specific details about their sourcing, manufacturing, and distribution practices.

Check the Ingredients: If you're shopping for food or personal care products, take a gander at the ingredients list. If it looks like a science experiment gone wrong, reconsider. Opt for stuff with natural ingredients. If you can't pronounce it, chances are it's not good for you or the planet. Look for shorter ingredient lists with recognizable components.

Social Media and Reviews: Nowadays, you can learn a lot about a brand just by scrolling through their social media. Check out customer feedback. If you see complaints about greenwashing (that's when a company pretends to be eco-friendly but really isn't), it's a red flag. Pay attention to how the company responds to criticism and whether they engage in meaningful conversations about sustainability.

Supporting Sustainable Brands

Now that we know how to spot eco-friendly products, let's chat about the importance of ethical consumerism. You might be asking, "Why should I care where my stuff comes from?" Here's the scoop:

When you buy from sustainable brands, you're not just making a purchase; you're making a statement. You're saying, "I care about the planet, and I want to support companies that do too!" Your dollars act like votes in the marketplace. The more we support ethical brands, the louder our message to other companies that sustainability matters. It's like a giant game of Monopoly, but instead of buying Boardwalk, you're investing in a cleaner planet. You're choosing to support businesses that prioritize environmental protection, fair labor practices, and community well-being.

Here's a fun fact: According to a Nielsen report, 66% of consumers are willing to shell out more for sustainable brands. That's a hefty chunk! It's like saying, "Hey, I'll pay a few extra bucks if it means I'm helping the Earth." And guess what? When more of us do this, companies start to take notice. They'll change their practices to match what we want. It's like being part of a movement, and who doesn't want to be part of something bigger than themselves? This demonstrates the growing consumer demand for sustainable products and the power of collective action.

Now, let's talk about supporting local businesses that prioritize sustainability. You know, those mom-and-pop shops that make your community special. Here's why this is crucial:

Boosting the Local Economy: Shopping local keeps money in your community. It's like watering a plant—it helps it grow! Plus, local businesses are more likely to source their materials sustainably and support other local initiatives. This creates a stronger, more resilient local economy.

Building Relationships: When you buy from local businesses, you often get to know the owners. They're your neighbors! You can chat with them about their practices and learn more about their commitment to sustainability. It's like having a personal tour guide for eco-friendly shopping. This fosters a sense of community and allows for greater transparency and accountability.

Reducing Carbon Footprint: Buying local means shorter shipping distances. Less transportation equals less pollution. It's like getting a two-for-one deal: you snag awesome products, and you help cut down your carbon footprint. Talk about a win-win! This reduces greenhouse gas emissions associated with long-distance transportation.

Finding Unique Products: Local businesses often offer unique, handcrafted items you won't find at big-box stores. You might stumble upon a local artisan who makes beautiful, sustainable goods. Plus, you can feel good knowing your purchase supports someone's dream. This adds character and diversity to your community while supporting local creativity and craftsmanship.

So, how do you find these local gems? Here are some tips:

Farmers' Markets: These are a treasure trove of local produce and products. Plus, you get to meet the farmers! It's like a mini-field trip every weekend. Farmers markets offer a direct connection to local food sources and support sustainable agriculture.

Community Events: Check out local fairs, festivals, and events. They often feature local vendors who prioritize sustainability. You might discover your new favorite go-to spot. These events provide a platform for local businesses to showcase their products and connect with the community.

Social Media: Follow local businesses on social media. They often post about their sustainable practices and any special events. Plus, it's a great way to stay in the loop! Social media can be a powerful tool for discovering and supporting local sustainable businesses.

Now, I know what you're thinking: "This all sounds great, but can I really make a difference?" Absolutely! Every little bit counts. When you choose to support sustainable brands, you're part of a larger movement. It's like being a superhero for the planet, one purchase at a time. Your choices send a powerful message to the marketplace and contribute to a more sustainable future.

Let's wrap this up with some actionable steps you can take today:

Start small: Next time you need to buy something, do a quick search for eco-friendly alternatives. You might be surprised at what you find! Even small swaps, like choosing reusable water bottles or shopping bags, can make a difference.

Make a list: Create a list of local businesses that prioritize sustainability. Keep it handy for your next shopping trip. This will make it easier to support local and sustainable businesses when you're out shopping.

Share the love: Talk to your friends and family about the importance of supporting sustainable brands. The more people who join the movement, the bigger the impact! Word-of-mouth marketing is a powerful tool for spreading awareness and encouraging others to make sustainable choices.

Challenge yourself: Set a goal to buy from at least one sustainable brand each month. Track your progress and see how it feels to make conscious choices. This can help you develop new habits and make sustainable shopping a regular part of your routine.

In conclusion, supporting sustainable brands is more than just a trendy choice; it's a powerful way to create positive change. You're not just buying products; you're voting for a brighter, cleaner future. So, let's roll up our sleeves and get to work. The planet needs us, and together, we can make a difference!

Let's break this down even further. Supporting sustainable brands is about being mindful of the choices we make. It's about understanding that every purchase has an impact. Think about it: when you buy something, you're not just getting a product; you're also supporting the practices behind it. You're supporting the way the product was made, the materials used, and the people involved in its production.

Take clothing, for example. Fast fashion is a huge culprit when it comes to environmental damage. It's all about quick trends and cheap prices, but at what cost? The workers are often underpaid, and the materials are usually low quality, leading to waste. Instead, consider brands that focus on sustainable practices. They might use organic cotton, recycled materials, or ethical labor practices. When you choose to buy from these brands, you're sending a message that you value quality and ethics over quantity. You're supporting fair wages, safe working conditions, and environmentally responsible manufacturing.

And let's not forget about the power of word-of-mouth. When you support sustainable brands, talk about it! Share your experiences on social media, recommend your favorite eco-friendly products to friends, and encourage others to join the movement. It's like a ripple effect. One person shares, then another, and before you know it, you've got a whole community buzzing about sustainability. This can create a powerful network of conscious consumers.

You know what's cool? Many sustainable brands are also super innovative. They're not just trying to be eco-friendly; they're rethinking how products are made and distributed. For instance, some companies are embracing a circular economy, where products are designed to be reused or recycled. It's a game-changer! Imagine a world where waste is minimized, and resources are used wisely. That's the kind of future we should all be rooting for. This approach reduces waste, minimizes resource depletion, and minimizes pollution.

Speaking of innovation, have you noticed how many brands are incorporating technology into their sustainability efforts? From apps that help you track your carbon footprint to companies using blockchain to ensure ethical sourcing, tech is playing a big role in making sustainability more accessible. It's exciting to see how creativity and technology can work hand-in-hand to solve some of our biggest challenges. These technological advancements can provide greater transparency and accountability within supply chains.

And let's chat about the younger generation for a sec. They're leading the charge when it comes to sustainability. Gen Z and millennials are more conscious about their purchases than ever before. They're demanding transparency from brands and aren't afraid to call out those who aren't living up to their promises. It's refreshing to see this shift in consumer behavior, and it's a powerful reminder that we all have a role to play in shaping the future. This increased awareness and activism among young people is driving significant change in the marketplace.

Now, let's not sugarcoat it—supporting sustainable brands isn't always easy. Sometimes, it can feel overwhelming. With so many options out there, it's tough to know where to start. But remember, every small step counts. You don't have to overhaul your entire lifestyle overnight. Just pick one or two changes to make, and go from there. Focus on making gradual, sustainable changes rather than trying to do everything at once.

For instance, if you're a coffee drinker, why not switch to a brand that prioritizes fair trade? Or if you're in the market for new cleaning supplies, look for eco-friendly options. Little by little, these changes add up. These small swaps can make a big difference in your overall environmental impact.

And hey, if you mess up and buy something that isn't sustainable? Don't beat yourself up about it. We're all human, and it's part of the learning process. The key is to stay informed and keep trying. It's important to be forgiving of yourself and to focus on making positive changes moving forward.

Oh, and let's not forget about the impact of our food choices. The food industry has a massive carbon footprint, and the way we eat can either help or hurt the planet. Consider incorporating more plant-based meals into your diet. You don't have to go full vegan, but even one meatless day a week can make a difference. Plus, you might discover some delicious new recipes along the way! Reducing your meat consumption can significantly reduce your environmental impact.

Now, if you're feeling a bit lost in the sea of sustainable brands, there are tons of resources out there to help you navigate. Websites and apps can help you find eco-friendly products, compare brands, and read reviews. It's like having a personal assistant for your shopping needs! These resources can provide valuable information about a brand's sustainability practices, certifications, and overall environmental impact.

And let's talk about the power of community. Joining local sustainability groups or online forums can be a great way to connect with like-minded folks. You can share tips, swap ideas, and even organize events to promote sustainable living. There's strength in numbers, and together, we can amplify our voices and make a bigger impact. Connecting with others who share your values can provide support, motivation, and inspiration.

So, as we wrap this up, remember that supporting sustainable brands is more than just a trend—it's a commitment to a better future. Every time you make a conscious choice, you're contributing to a larger movement. It's about creating a world where people and the planet can thrive together.

It's about voting with your wallet and supporting companies that are making a positive difference.

Chapter 9

Gardening for the Planet

Alright, let's dig into this! No pun intended. Gardening isn't just about pretty flowers or a fresh salad; it's a game-changer for our planet. Growing your own food has benefits that go beyond just saving a few bucks at the grocery store. It's about sustainability, health, and supporting our local ecosystems. So grab your gloves and let's get started! Think about the satisfaction of harvesting your own vegetables, knowing exactly where they came from and how they were grown.

First off, let's talk about the benefits of growing your own food. Imagine biting into a juicy tomato that you nurtured from a tiny seed. There's something magical about that, right? You're not just eating; you're connecting with the earth. Here are some solid perks:

1. **Freshness**: Homegrown food tastes better. It's like comparing a home-cooked meal to a frozen dinner. You can't beat that freshness! The flavors are more vibrant and intense when food is picked at its peak ripeness and eaten soon after.

2. **Nutrition**: You control what goes into your soil and your plants. No pesticides or mystery ingredients. Just pure, healthy goodness. You can choose organic methods and ensure that your food is free from harmful chemicals and additives.

3. **Sustainability**: You're reducing your carbon footprint. Think about it: no plastic packaging, no long-distance transport. Just you, your seeds, and a little patch of earth. Growing your own food reduces the energy and resources required for commercial agriculture, packaging, and transportation.

4. **Mental Health**: Gardening is therapeutic. It's like a mini-vacation from life's chaos. Getting your hands dirty can reduce stress and boost your mood. Seriously, it's science! Studies have shown that gardening can lower cortisol levels (a stress hormone) and promote feelings of well-being.

Now, you might be thinking, "But I don't have a yard!" Don't sweat it. You can start a garden in even the tiniest of spaces. Here's how to kick off your gardening journey, no matter where you live:

1. **Choose Your Space**: Got a balcony? A windowsill? Even a countertop can work. Look for areas that get plenty of sunlight. Most vegetables need at least six hours of sunlight per day.

2. **Start Small**: Don't overwhelm yourself with a massive garden. Start with a few pots of herbs or a couple of tomato plants. It's like dipping your toes in the water before diving in. This allows you to learn the basics and gain confidence before expanding your garden.

3. **Use Containers**: Container gardening is your friend. You can use old buckets, pots, or even repurposed items like tin cans. Just make sure there are drainage holes! Proper drainage prevents water from accumulating at the bottom of the container, which can lead to root rot.

4. **Select Easy Plants**: Herbs like basil, mint, and chives are super forgiving. They'll grow like weeds and are great for cooking! If you want something more substantial, try cherry tomatoes or lettuce. These plants are relatively easy to grow and require minimal maintenance.

5. **Water Wisely**: Don't drown your plants! Overwatering is a common rookie mistake. Check the soil—if it's dry an inch down, it's time to water. Overwatering can lead to root rot and other plant diseases.

6. **Learn as You Go**: Gardening is a learning experience. You'll make mistakes, but that's part of the fun. Embrace it! Don't be afraid to experiment and try new things. There are plenty of resources available online and at local libraries to help you learn.

Now, let's dig a little deeper (pun absolutely intended) into the role of native plants in supporting local ecosystems. Native plants are like the unsung heroes of our environment. They've adapted to local conditions over thousands of years and provide essential benefits:

1. **Biodiversity**: Native plants attract local wildlife, including pollinators like bees and butterflies. More diversity means a healthier ecosystem. These pollinators are crucial for the reproduction of many plants, including many of our food crops.

2. **Soil Health**: These plants help prevent erosion and improve soil quality. They're like nature's little gardeners, keeping everything in check. Their root systems help to bind the soil together, preventing it from being washed away by rain or wind.

3. **Water Conservation**: Native plants require less water once established. They're already adapted to your local climate, making them super low-maintenance. This is especially important in areas with limited water resources.

4. **Support Local Species**: Many birds and insects rely on native plants for food and shelter. By planting them, you're creating a haven for local wildlife. This helps to maintain the delicate balance of the local ecosystem.

So, how do you incorporate native plants into your gardening plans? Here's a quick checklist:

- **Research Local Species**: Check out local nurseries or extension services for native plant lists. You want plants that thrive in your area. These resources can provide valuable information about the best native plants for your specific region.

- **Create a Native Garden Bed**: Dedicate a small area of your garden to native plants. Mix and match different species for a vibrant look. This creates a designated space for native plants to thrive and attract local wildlife.

- **Join Community Efforts**: Many communities have initiatives to plant native species. Get involved and meet fellow eco-warriors! These initiatives offer a great opportunity to learn more about native plants and contribute to local conservation efforts.

- **Educate Others**: Share what you've learned about native plants with friends and family. The more people know, the more we can all contribute to a healthier planet. Spreading awareness about the benefits of native plants can encourage more people to incorporate them into their own gardens.

Now, you might be wondering, "How do I find the time for all this?" Well, here's the deal: gardening doesn't have to be a full-time job. Start with just a few minutes each day. Water your plants, pull a weed, or simply enjoy the view. Over time, those little moments add up. Even just 15-20 minutes a day can make a big difference in the health and productivity of your garden.

And let's not forget about the community aspect of gardening. It's a fantastic way to connect with others who share your passion for the planet. Join a local gardening club or community garden. You'll learn new skills, swap seeds, and maybe even make some lifelong friends. Community gardens also provide opportunities to grow food for those in need.

In conclusion, gardening is more than just a hobby; it's a powerful way to contribute to a sustainable future. You're not just growing food; you're nurturing the planet. So whether you have a sprawling backyard or a tiny apartment, there's always room for a little green. You're creating a positive impact on the environment, your health, and your community.

Now, get out there, dig in, and remember: every seed you plant is a step toward a brighter, greener future. Let's grow together!

Chapter 10

Advocating for Change

Alright, folks, let's dive into the good stuff. You're a teen with passion and energy, and guess what? You have the POWER to create real change. Seriously, the world needs your voice, your ideas, and your drive. Youth activism is like a fresh breeze blowing through a stuffy room—it shakes things up, gets people talking, and can even rattle some cages. So, let's break this down into three juicy parts: the power of youth activism, how to effectively communicate your message, and how to engage with local leaders and policymakers. Buckle up, because we're about to make some waves! Consider the impact you can have not just on your local community, but on a global scale as well.

First off, let's chat about the power of youth activism. Remember the last time you felt like you could take on the world? Maybe it was during a heated debate in history class or when you rallied your friends to clean up the park. That feeling? That's what activism is all about! You see, young people have a unique perspective on the world. You're not weighed down by decades of "this is how it's always been." You see the problems clearly and are often more willing to challenge the status quo. You have a fresh perspective and the energy to create positive change.

Take Greta Thunberg, for example. She's like the rockstar of climate activism. She started by skipping school to protest outside the Swedish parliament. Can you believe that? One girl, a sign, and a whole movement was born! That's the kind of impact you can have. Your voice matters, and it can inspire others to join. So, whether it's organizing a local clean-up or starting a social media campaign, remember: your actions can ignite change.

Now, let's shift gears a bit and talk about how to effectively communicate your message. This is where things can get a little tricky, but don't sweat it! Here are some practical tips to help you get your point across:

1. **Know your audience.** Who are you trying to reach? Are they your classmates, local leaders, or maybe even your parents? Tailor your message to resonate with them. Consider their interests, concerns, and level of understanding of the issues.

2. **Use stories.** People connect with stories. Share your experiences or those of others who have been affected by environmental issues. It's like the difference between reading a textbook and watching a gripping movie—one's way more engaging! Personal stories can create an emotional connection and make the issue more relatable.

3. **Keep it simple.** Don't drown your audience in jargon. You want to inspire, not confuse. Use clear,

straightforward language. Think of it as explaining your favorite video game to your grandma—keep it relatable! Avoid using technical terms or complex statistics unless you are sure your audience will understand them.

4. **Call to action.** Don't just inform; inspire action! Tell your audience what they can do to help. Whether it's signing a petition or joining a community event, give them a clear path forward. Make it easy for people to get involved by providing specific steps they can take.

5. **Use social media.** Platforms like Instagram, TikTok, and Twitter are your best friends. Use them to share your message, connect with others, and rally support. Just remember to keep it positive and constructive—no one likes a troll! Use relevant hashtags, create engaging content, and interact with your followers to build a strong online presence.

Now that you've got the communication down, let's talk about engaging with local leaders and policymakers. This might sound intimidating, but it's easier than you think. Here's a little secret: they want to hear from you! Your voice can influence decisions that affect your community and the planet. Here's how to make it happen:

Research your local leaders. Know who they are and what they stand for. This will help you tailor your message and find common ground. Learn about their positions on environmental issues and their voting records.

Schedule a meeting. Don't be shy! Reach out and ask for a few minutes of their time. Be polite, and don't forget to express your appreciation for their work. Prepare a concise and respectful request for a meeting.

Prepare your pitch. Before the meeting, outline what you want to say. Keep it focused and concise. You might even want to practice with a friend or family member. This will help you feel more confident and prepared during the meeting.

Share your story. Use your personal experiences to illustrate why these issues matter. Remember, you're not just another voice; you're a passionate advocate for change! Sharing personal anecdotes can make your message more impactful and memorable.

Follow up. After your meeting, send a thank-you note. It shows you're serious and keeps the lines of communication open for future conversations. This demonstrates professionalism and maintains a positive relationship with the leaders.

Now, I know what you might be thinking: "This sounds great, but what if I fail?" Here's the thing: failure is just a stepping stone to success. Every great leader has faced setbacks. It's how you bounce back that counts. Think of it like riding a bike—sure, you might fall a few times, but eventually, you'll be cruising down the street like a pro. Don't let fear of failure hold you back from taking action.

And hey, let's not forget about the power of community. Surround yourself with like-minded folks who share your passion for the planet. Join local environmental groups, start a club at school, or even organize a virtual meetup. When you come together, you amplify your impact. Plus, it's way more fun to tackle these challenges with friends! Working together can provide support, motivation, and a sense of shared purpose.

So, what's the takeaway here? You have the power to advocate for change. Use your voice, communicate effectively, and engage with your local leaders. Don't underestimate your potential. Remember, every big movement starts with a single step. Whether you're leading a march, writing a letter to your mayor, or simply having a conversation with a friend, you're making a difference. Every action, no matter how small, contributes to the larger goal of creating a sustainable future.

Your passion and dedication can inspire others to join the movement and create a positive impact on the world.

Chapter 11

Climate Justice

Alright, let's jump into the sizzling topic of climate justice. I mean, it's hotter than a jalapeño in July, right? You might be scratching your head, thinking, "What's the deal? How's climate change tied to social justice?" Well, grab a drink, kick back, and let's break it down. Climate justice recognizes that climate change doesn't affect everyone equally, and it's crucial to understand these disparities to create effective solutions.

So, here's the scoop: climate change isn't just about the Earth getting a fever. It's about people—real, living, breathing folks. And guess what? It's often the marginalized communities that get the short end of the stick. Picture this: if climate change were a game of dodgeball, the folks in the back—the ones already juggling economic hardships—are the first to get whacked. They're the ones stuck living next to factories spewing pollution, navigating food deserts, and trying to cope with wild weather. It's like being trapped in a horror flick, and the monster? Yep, it's climate change. These communities often lack the resources and political power to protect themselves from the worst impacts of climate change.

Now, let's chat about the intersection of climate change and social equity. Not everyone's hit the same way. Wealthy neighborhoods? They've got the resources to weather the storm (literally). They can afford to evacuate, rebuild, and adapt. Meanwhile, low-income areas are often left high and dry, lacking the basic infrastructure to deal with climate impacts. They may lack access to adequate housing, healthcare, and other essential services. It's a tough pill to swallow, but it's the reality. If we want to save the planet, we've to lift everyone up—not just those who can slap solar panels on their roofs or drive electric cars. True climate solutions must address the root causes of social and economic inequality.

Speaking of numbers, did you know that, according to the EPA, people of color are more likely to live in areas with higher pollution levels? Yeah, it's a real downer. Studies show that Black and Hispanic communities often deal with more exposure to toxic waste and air pollution. This isn't just about breathing bad air; it's about health risks, economic instability, and overall quality of life. It's a domino effect that messes with everything from job opportunities to educational resources. This disproportionate exposure to environmental hazards is known as environmental racism.

So, how do we turn the tide? How can we back climate justice initiatives and actually make a difference? Here are some steps you can take:

First up, educate yourself and others. Knowledge is like gold, folks! Dive into books, articles, and documentaries that shine a light on the link between climate change and social justice. Share what you find with your crew. You might just spark a convo that leads to some real action. Understanding the complexities of climate justice is the first step towards making a positive change.

Next, support local organizations. Find groups that focus on climate justice. Many of them work at the grassroots level, advocating for policies that protect marginalized communities. Whether it's donating cash, volunteering your time, or just spreading the word, every little bit helps. These organizations often provide direct support to communities impacted by climate change and advocate for more equitable policies.

Then there's advocacy for policy change. Get involved in local politics! Attend town hall meetings, write to your reps, and demand policies that prioritize climate justice. Your voice matters! You can be the change-maker in your community. This includes advocating for policies that reduce pollution in marginalized communities, invest in renewable energy in low-income areas, and ensure access to clean water and sanitation for all.

Also, practice inclusive environmentalism. When you're getting involved in environmental initiatives, make sure to include voices from marginalized communities. This isn't just about planting trees; it's about creating a movement that represents everyone. Think of it like a potluck dinner—everyone should bring a dish to the table. Diverse perspectives and experiences are essential for creating effective and equitable solutions.

And don't forget to be mindful of your own impact. While we're all about saving the planet, it's crucial to recognize our own privilege. Reflect on how your lifestyle choices might affect others. Can you cut down on waste? Support ethical brands? Every action counts, big or small. Consider the environmental impact of your consumption habits and strive to make more sustainable choices.

Now, let me get a bit personal. I remember when I first dipped my toes into environmental activism. I was fired up, ready to change the world, but quickly realized I was missing a key piece. I was all about recycling and reducing my carbon footprint, but I wasn't thinking about how those efforts intersected with social equity. It was a wake-up call. I started attending community meetings and listening to the stories of those directly impacted by environmental issues. That's when I got it—climate justice isn't just a side note; it's central to the fight against climate change. Understanding the human dimension of climate change is essential for creating effective and just solutions.

Think about it: if we want a sustainable future, we need to ensure that everyone has a seat at the table. It's like trying to bake a cake without all the ingredients. You might end up with something that looks nice on the outside, but it won't taste right. We need a recipe that includes diverse voices, perspectives, and experiences. This ensures that solutions are effective, equitable, and address the needs of all communities.

And here's the kicker—climate justice isn't just a moral imperative; it's also an economic one. Investing in marginalized communities can lead to job creation, better health outcomes, and stronger local economies. It's a win-win! By supporting initiatives that promote sustainability and equity, we're not just saving the planet; we're building a brighter future for everyone. This includes investing in green jobs in underserved communities and promoting economic development that is both sustainable and equitable.

As we wrap this up, let's take a sec to reflect on what we've learned. Climate justice is about understanding the connections between environmental issues and social equity. It's about recognizing that marginalized communities often face the greatest risks and taking action to support them. It's about creating a world where everyone has the right to a healthy environment.

So, what's your next move? Here's a little challenge: pick one of those practical steps I mentioned and put it into action this week. Whether it's diving into some reading on climate justice or volunteering with a local group, take that leap. You've got the power to make a difference, and every small action can lead to BIG changes. Start small, but start now.

Remember, the fight for climate justice is ongoing, and it requires all of us. Together, we can create a sustainable and equitable future for everyone. Now, go out there and be the change you want to see in the world! Your actions can inspire others and contribute to a more just and sustainable future.

But wait, let's dig a little deeper. What does it really mean to stand up for climate justice? It's about more than just awareness. It's about building a movement that's loud and proud, that demands change at every level—from local communities to global platforms. It's about making sure that the voices of those most affected by climate change are heard and prioritized. It's about dismantling systems of oppression that contribute to environmental injustice.

Imagine living in a neighborhood where the air is thick with smog, where kids can't play outside without coughing, and where the nearest grocery store is miles away. That's the reality for many marginalized communities. They're not just facing climate change; they're dealing with systemic inequalities that make everything worse. It's like they're stuck in a never-ending game of whack-a-mole, where every time they tackle one issue, another one pops up. These inequalities can include poverty, lack of access to healthcare, and limited political representation.

Now, let's get real about the role of corporations in this whole mess. Big companies often prioritize profit over people. They set up shop in low-income areas, knowing they can get away with polluting because the residents don't have the resources to fight back. It's a classic case of environmental racism. And the kicker? Those same communities often don't see the benefits of the economic activity that's happening right in their backyard. It's a raw deal, and it's got to change. These corporations have a responsibility to operate in a sustainable and equitable manner.

So, how do we hold these corporations accountable? One way is through consumer power. When we choose to support businesses that prioritize sustainability and social equity, we send a message that we won't stand for environmental injustice. We can use our wallets to vote for the kind of world we want to see. Plus, it's a great conversation starter—"Hey, did you hear about that company that's actually making a difference?" You'd be surprised how quickly a little chat can snowball into something bigger. This can encourage companies to adopt more sustainable and ethical practices.

And don't underestimate the power of grassroots movements. History's shown us that when people come together, change happens. Think about the civil rights movement, the women's suffrage movement, or even recent climate strikes led by young activists. These movements are proof that collective action can turn the tide. It's about organizing, rallying, and making some noise until the powers that be can't ignore us anymore. Grassroots movements can create significant political pressure and bring about meaningful change.

Now, let's talk about the role of technology in this fight. Yeah, tech can be a double-edged sword. On one hand, it's responsible for a lot of the environmental mess we're in—hello, fossil fuels and e-waste. But on the flip side, technology can also be a powerful tool for change. Renewable energy sources, electric vehicles, and smart technology can help us reduce our carbon footprint. Plus, social media is a game changer for spreading awareness and mobilizing people. Just look at how quickly movements can gain traction online. It's like a wildfire of awareness, spreading faster than you can say "climate justice." Technology can be used to monitor pollution levels, track environmental impacts, and connect activists across the globe.

But here's the thing: we can't rely solely on technology to save us. We need to pair it with grassroots activism and policy change. It's a team effort, folks. We've to push for laws that protect our environment while also ensuring that marginalized communities are at the forefront of these conversations. It's about creating a holistic approach to climate justice—one that acknowledges the interconnectedness of social and environmental issues. This requires collaboration between scientists, policymakers, community leaders, and activists.

And let's not forget about the power of storytelling. Sharing personal stories can be a powerful way to highlight the impacts of climate change on individuals and communities. It humanizes the issue and makes it relatable. Think about it: when you hear someone's story about how climate change has affected their life, it hits differently than just hearing stats and figures. It's about making that emotional connection. Stories can inspire empathy, build understanding, and motivate action.

So, as we dive deeper into this, let's challenge ourselves to think about our own stories. What brought you to this conversation? What experiences have shaped your views on climate justice? Reflecting on our own journeys can help us connect with others and build a stronger movement. It can also help us identify our own biases and privileges.

Alright, let's get back to action. We've talked about a lot of different ways to get involved, but let's break it down into bite-sized pieces. Here's a quick checklist for you:

1. **Read up**: Grab a book or watch a documentary on climate justice. You could start with something like "This Changes Everything" by Naomi Klein or "The True Cost" to get the wheels turning. These resources can provide a deeper understanding of the issues and inspire you to take action.

2. **Join a local group**: Find an organization in your area that focuses on climate justice. Whether it's a community garden, a local advocacy group, getting involved can make a huge difference. Connecting with others who share your passion can provide support and motivation.

3. **Speak up**: Use your voice! Whether it's on social media or at community meetings, don't be afraid to share your thoughts on climate justice. You never know who might be listening. Your voice can inspire others to get involved and make a difference.

4. **Support minority-owned businesses**: Make a conscious effort to shop at businesses that prioritize sustainability and equity. It's a small change that can have a big impact. This supports local economies and promotes ethical business practices.

5. **Volunteer**: Find opportunities to lend a hand in your community. Whether it's cleaning up a local park or helping out at a food bank, every bit helps. Volunteering can provide direct support to communities in need and help to address environmental injustices.

6. **Engage with your elected officials**: Write letters, make phone calls, or even set up meetings to discuss climate justice issues with your representatives. Let them know it

matters to you. This is a powerful way to advocate for policy change and hold elected officials accountable.

7. **Spread the word**: Talk to your friends and family about climate justice. Share articles, post on social media, or even host a movie night to get the conversation going. Raising awareness is essential for building a broader movement for change.

8. **Reflect on your own lifestyle**: Think about how your choices impact the environment and marginalized communities. Can you make any changes to be more sustainable? This includes considering your consumption habits, transportation choices, and energy usage.

And here's a thought—what if we all committed to one action a week? Imagine the ripple effect! Small changes can lead to big waves. It's all about creating a culture of awareness and action. Consistent action, even in small ways, can create significant change over time.

As we wrap this up, remember: climate justice is about more than just the environment. It's about people. It's about recognizing the interconnectedness of our struggles and working together to build a better future.

Chapter 12

Eco-Friendly Technology

You ever look at your phone and wonder if it's a blessing or a curse? I mean, sure, it connects you with friends, helps you navigate to the nearest taco truck, and lets you binge-watch your favorite shows. But what about the impact on the planet? Let's dive into the world of eco-friendly technology, where innovation meets sustainability. Trust me; it's not just about saving the whales. It's about saving ourselves, too. Consider the resources that go into manufacturing our devices, from the mining of raw materials to the energy used in production and transportation.

First off, let's talk INNOVATIONS that PROMOTE SUSTAINABILITY. The tech world is buzzing with gadgets that are designed not just to make our lives easier but also to be kinder to Mother Earth. For instance, have you heard of solar-powered chargers? They're like the sun's little gift to your phone. Just plop that bad boy in the sun, and you're good to go. No more hunting for an outlet in a crowded café or worrying about battery life while you're out hiking. This reduces reliance on traditional electricity sources and harnesses renewable energy.

Then there are smart home devices. These nifty gadgets can help you monitor and control your energy usage. Think of smart thermostats that learn your schedule and adjust the temperature accordingly. You'll save energy and money, and you can feel like a tech-savvy eco-warrior at the same time. It's like having your cake and eating it too, but the cake is made of recycled materials. These devices can also help you track water usage, control lighting, and manage other household systems to maximize efficiency.

Now, let's not forget about electric vehicles (EVs). If you haven't taken a ride in one yet, you're missing out. These cars are the future of transportation, reducing carbon emissions and giving you that smooth, quiet ride that makes you feel like you're gliding on air. Plus, they come with some serious tech features that make driving feel like a scene from a sci-fi movie. I mean, who wouldn't want to feel like they're in "Back to the Future"? EVs contribute to cleaner air and reduced reliance on fossil fuels.

But here's the kicker: the role of tech in reducing carbon footprints goes beyond just cool gadgets. It's about how we use these innovations to make a real difference. Every time you choose an electric car over a gas guzzler, you're making a statement. Every time you opt for energy-efficient appliances, you're saying, "Hey, I care about this planet!" These choices send a message to manufacturers and policymakers about the importance of sustainability.

Now, you might be thinking, "But how do I know if my gadgets are actually eco-friendly?" Great question! Evaluating the environmental impact of gadgets and devices is crucial. Start by looking for ENERGY STAR ratings. These labels are like gold stars for appliances that meet energy efficiency guidelines set by the U.S. Environmental Protection Agency. It's like a badge of honor for being green. This label indicates that the product meets specific energy-saving criteria.

Another tip? Check the materials. Are they recyclable? Can they be repaired instead of tossed? If a device is designed to break down after a couple of years, it's time to rethink that purchase. Go for companies that prioritize sustainability in their manufacturing processes. Look for products made from recycled materials, with minimal packaging, and designed for durability and repairability.

And let's talk about the life cycle of a product. It's not just about how it works when you buy it. Consider what happens when it's time to say goodbye. Does it end up in a landfill, or can it be recycled or repurposed? Many companies are now offering take-back programs, which means they'll take your old tech and recycle it properly. It's like a second chance for your gadgets, and it keeps them out of landfills. This reduces electronic waste and recovers valuable materials.

Now, let's not sugarcoat it. The tech industry has a long way to go. With all the advancements, there's still a hefty carbon footprint associated with manufacturing and shipping these devices. But the good news? Many companies are stepping up. They're investing in renewable energy, using sustainable materials, and creating products that are designed to last. So, when you're out shopping, keep an eye out for those brands that are making a genuine effort. Look for companies that are transparent about their environmental practices and committed to reducing their impact.

Here's a fun little challenge for you: Next time you're in the market for a new gadget, make a list of eco-friendly features you want. Maybe it's a product made from recycled materials, or perhaps it's something that's energy-efficient. Hold yourself accountable. You'd be surprised at how much power you have as a consumer. This can help you make more informed purchasing decisions and support companies that are doing their part for the environment.

Let's wrap this up with a little reflection. Technology is a tool, and like any tool, it can be used for good or for bad. The key is to be intentional about how we use it. If we can harness the power of innovation to create a more sustainable future, why wouldn't we? By choosing eco-friendly tech, we can reduce our environmental impact and contribute to a more sustainable future.

So, next time you reach for that shiny new gadget, ask yourself: Is it just a pretty face, or is it making a difference?

Remember, every small choice adds up. Together, we can turn the tide and create a planet that's not just surviving but THRIVING. We can use technology to create a more sustainable and equitable world for everyone.

In conclusion, embrace the eco-friendly tech revolution. Dive into those innovations that promote sustainability, leverage technology to reduce your carbon footprint, and always evaluate the environmental impact of your gadgets.

It's time to become a savvy consumer and a responsible global citizen. By making informed choices about the technology we use, we can contribute to a more sustainable future for ourselves and the planet.

Chapter 13

Sustainable Fashion

Alright, let's jump into something that's hotter than a jalapeño on a summer day—Sustainable Fashion. Now, before you roll your eyes and think, "Oh great, another lecture on how to dress like a hippie," just hang tight. This isn't just about looking like you stepped out of a nature documentary. It's about feeling good, looking sharp, and doing right by Mother Earth. So grab that vintage tee you snagged at a thrift store, and let's get into the nitty-gritty! Consider the entire lifecycle of clothing, from the sourcing of raw materials to manufacturing, transportation, use, and disposal.

First up, let's tackle the elephant in the room: fast fashion. If you think fast food's bad for your waistline, wait till you hear about the damage fast fashion does to our planet. This industry cranks out millions of garments every year—most of which are worn once or twice before they're tossed aside like yesterday's pizza. Can you believe that the fashion world is responsible for around 10% of global carbon emissions? Yep, that's a jaw-dropper! Every time you snag a cheap shirt, somewhere a polar bear is shedding a tear. Fast fashion doesn't just mess with the climate; it also creates a ton of waste. Those trendy outfits you bought on a whim? They're headed straight to landfills, adding to the staggering 92 million tons of textile waste we churn out annually. Crazy, right? This waste contributes to pollution, takes up valuable landfill space, and often contains harmful chemicals.

So what's the game plan? Let's build a sustainable wardrobe. It sounds daunting, but trust me, it can be a fun ride! Here's a quick list to get you started:

- **Quality Over Quantity**: Think about it—would you rather have a closet full of cheap stuff that falls apart after a wash or a few solid pieces that last? Investing in high-quality items is like choosing a trusty pair of boots over flimsy flip-flops. A good pair of jeans can stick with you for years, while the cheap ones might give up on you before your next birthday. This reduces the need to constantly buy new clothes, which saves resources and reduces waste.

- **Timeless Styles**: Go for classic cuts and colors that won't go out of style. A little black dress or a sharp blazer can be your best friends. Mix and match these staples for all sorts of occasions. You'll look like you've got it all together when really, you're just playing it smart! This helps to avoid the constant cycle of buying into fleeting trends.

- **Sustainable Brands**: Support brands that care about the planet. There are tons of companies out there making clothes from recycled materials or using eco-friendly production methods. A little research goes a long way—your wallet and the Earth will thank you! Look for certifications like Fair Trade, GOTS (Global Organic Textile Standard), and Bluesign.

- **Capsule Wardrobe**: Ever heard of a capsule wardrobe? It's like Marie Kondo meets fashion! You pick a small collection of versatile pieces that you love and can wear in different ways. This not only saves you time in the morning but also keeps your closet from looking like a tornado hit it. This encourages mindful consumption and reduces clutter.

Now, I know what you're thinking: "But I can't afford all that fancy stuff!" Slow down there, my friend. That's where thrifting and clothing swaps come into play. These aren't just trendy terms; they're your new best pals in the fight against fast fashion. These options make sustainable fashion more accessible and affordable.

Let's break it down:

- **Thrifting**: Thrifting is like a treasure hunt. Who doesn't love an adventure? You can score unique pieces that no one else has, and it's usually way cheaper than buying new. Plus, you're giving clothes a second life instead of letting them rot in a landfill. Win-win! Just remember to wash everything when you bring it home—trust me, some thrift stores have a unique smell that you don't want lingering in your closet.

- **Clothing Swaps**: Organize a clothing swap with your friends. It's like a potluck but for clothes! Everyone brings items they don't wear anymore, and you get to pick out new-to-you pieces. It's a fun way to refresh your wardrobe without spending a dime. Plus, it's a great excuse to hang out and gossip about that one friend who always shows up late.

Feeling a bit overwhelmed? "Where do I even start?" you ask. Here's a little challenge for you:

- **Pick one day this week to go through your closet.** Pull out anything you haven't worn in the last year. If it doesn't spark joy (thanks, Marie Kondo), think about donating it or tossing it in the swap pile.

- **Set a budget for thrifting.** Maybe it's $5. Maybe it's $50. Whatever it is, stick to it! You'll be surprised at what you can find without breaking the bank.

- **Try to add at least one sustainable piece** to your wardrobe this month. It could be a t-shirt made from organic cotton or a pair of shoes from a brand that gives back to the community.

Let's not forget the role social media plays in this whole sustainable fashion movement. Platforms like Instagram and TikTok are overflowing with influencers and creators who are all about eco-friendly fashion. Follow a few of them for inspiration, tips, and even DIY ideas. You might just stumble upon a new favorite outfit or a creative way to upcycle something you already own! Social media can be a powerful tool for spreading awareness and connecting with others who are passionate about sustainable fashion.

And here's a little secret: Sustainable fashion isn't just about what you wear; it's about how you feel in it. When you know you're making choices that benefit the planet, it gives you a sense of purpose. You walk a little taller, your smile shines a little brighter, and you might even inspire someone else to think twice before they grab that fast fashion dress. It's about feeling good about your choices and contributing to a positive change.

In wrapping this up, sustainable fashion is about making smarter choices, building a wardrobe that reflects your values, and having a blast while doing it. The next time you're tempted to buy that cheap shirt, ask yourself: "Is this worth the environmental cost?" You've got the power to change the narrative, one outfit at a time. So go on, strut your sustainable stuff, and let's save the planet—one thrift store at a time!

But wait, there's more! Let's dive deeper into the nitty-gritty of sustainable fashion, shall we? We've only scratched the surface here.

Let's chat about the fabrics. You know, what your clothes are made of matters. Ever heard of organic cotton? It's like the cool cousin of regular cotton. Grown without harmful pesticides and fertilizers, it's way better for the planet and your skin. And then there's Tencel—made from wood pulp, it's biodegradable and super soft. It's like wearing a hug! Plus, there's recycled polyester, which takes plastic bottles and turns them into stylish threads. Talk about a win for the environment! Different fabrics have different environmental impacts, so it's important to choose wisely.

Then there's the whole issue of labor. Fast fashion often relies on sweatshops where workers are paid peanuts and work in terrible conditions. Supporting sustainable brands means backing companies that treat their workers fairly. You want to wear clothes that make you feel good inside and out, right? Knowing the people who made your clothes are treated well is a huge part of that. Look for Fair Trade certifications to ensure that workers are paid a living wage and work in safe conditions.

Now, let's talk about the lifecycle of clothing. Ever thought about where your clothes go when you're done with them? Most people toss them in the trash without a second thought. But there are so many better options! You can donate them, sell them, or even repurpose them. Got an old pair of jeans? Turn them into a trendy bag or some cool patches. The possibilities are endless! This reduces waste and conserves resources.

And here's a thought: how about learning to sew? Sounds intimidating, but it's actually pretty fun! You could mend your favorite pair of jeans instead of tossing them. Or, get creative and turn an old shirt into a stylish crop top. There's something super satisfying about wearing something you made yourself. This extends the life of your clothes and reduces the need to buy new items.

Let's not forget about the impact of fashion shows and the industry's obsession with trends. Ever notice how one minute it's all about skinny jeans, and the next, it's wide-leg trousers? It's a whirlwind! This constant push for newness fuels fast fashion and encourages overconsumption. But you can fight back! Stick to your style and wear what you love, regardless of what's "in" at the moment. This helps to break the cycle of constant consumption and reduces waste.

Now, how about the idea of renting clothes? It's like borrowing your best friend's outfit for a night out, but on a larger scale. There are services out there that let you rent high-end pieces for special occasions. You get to rock that designer dress without the hefty price tag or the guilt of it sitting in your closet forever. Plus, it's a fantastic way to try out new styles without commitment. This reduces the demand for new clothing production and minimizes textile waste.

And hey, let's sprinkle in some humor here. Ever seen someone walk into a party wearing the same outfit as you?

Awkward, right? With sustainable fashion, you're more likely to rock unique pieces that no one else has. Thrift stores are like a fashion roulette—who knows what gem you'll find? Embracing unique and pre-owned clothing can help you express your individuality and avoid fashion faux pas.

But, real talk—sustainable fashion isn't just a trend; it's a movement. It's about changing the way we think about our clothes and the impact they have on the world. It's about creating a culture where we value quality over quantity, where we appreciate the stories behind our garments, and where we stand up for the planet and the people who inhabit it. It's about creating a more ethical and environmentally responsible fashion industry.

So, what's your role in all this? Start conversations! Share what you've learned with your friends and family. The more people know about sustainable fashion, the bigger the impact we can make. You could even start a blog or an Instagram account dedicated to your sustainable journey. Who knows? You might inspire others to hop on the bandwagon! Sharing information and inspiring others is crucial for growing the sustainable fashion movement.

And let's not forget about the power of community. Join local groups focused on sustainable fashion or participate in events like clothing swaps and upcycling workshops. You'll meet like-minded folks, share ideas, and have a blast while doing it. Plus, it's always nice to know you're not in this alone. Connecting with others can provide support, motivation, and a sense of belonging.

In the end, sustainable fashion is about making choices that align with your values. It's about creating a wardrobe that tells your story while being kind to the planet. So the next time you're out shopping, take a moment to think about the bigger picture. Ask yourself if that fast fashion item is worth it or if you could find something more sustainable. Conscious consumerism is a powerful tool for driving change in the fashion industry.

Let's wrap this up with a little pep talk. You've got the power to make a difference, one outfit at a time. Embrace the adventure of sustainable fashion. It's not just about what you wear; it's about how you wear it and the impact you have on the world around you. So go ahead, strut your sustainable stuff, and let's save the planet—one thrift store at a time! Your choices can inspire others and contribute to a more sustainable future for the fashion industry.

And remember, every small step counts. Whether it's thrifting, swapping, or just being more mindful about your purchases, you're making a difference. So let's keep this momentum going. Who's with me? By working together, we can create a more sustainable and ethical fashion future.

Chapter 14

Mindfulness and Sustainability

Alright, let's dive into a topic that's not just a buzzword but a game-changer: MINDFULNESS and SUSTAINABILITY. Now, before you roll your eyes and think, "Oh great, another self-help session," let me assure you, this isn't your average run-of-the-mill chat. We're talking about a powerful connection that can shift how we live and interact with our planet. So grab a comfy seat, maybe a cup of tea (or coffee, no judgment here), and let's get into it. Think about how often we go through our days on autopilot, not fully present in our experiences.

First off, what's the deal with mindfulness? At its core, mindfulness is all about being present. It's like when you're munching on a slice of pizza and you actually pay attention to the taste, the texture, and the gooey cheese—rather than scrolling through your phone or zoning out. You know what I mean? When you're fully engaged, you start noticing things you might usually overlook. Now, imagine applying that same level of attention to your relationship with the environment. It's like turning on a light in a dark room. Suddenly, you see the mess, the beauty, and the potential for change. This awareness is the foundation of a sustainable lifestyle.

Now, here's where it gets juicy: being eco-conscious isn't just about recycling or using reusable bags. It's about cultivating a SUSTAINABLE MINDSET. This means shifting your perspective from "What can I get?" to "What can I give back?" It's like that classic saying: "You reap what you sow." If you're sowing awareness and intentionality, you're going to reap a healthier planet. This mindset encourages us to consider the long-term consequences of our actions and to prioritize the well-being of the planet.

So how do we cultivate this sustainable mindset? Well, it's not as complicated as it sounds. Here's a nifty little list to get you started:

1. **Start Small**: You don't have to overhaul your entire life overnight. Pick one thing to focus on. Maybe it's reducing your plastic use or being more mindful about your food choices. Every little bit counts! Small, consistent changes are more sustainable than trying to do everything at once.

2. **Educate Yourself**: Knowledge is power, folks. Read up on sustainability, watch documentaries, or listen to podcasts. The more you know, the more you can make informed choices. Plus, you'll sound super smart at parties. Understanding the science behind environmental issues can motivate you to make changes in your own life.

3. **Practice Gratitude**: Seriously, take a moment each day to appreciate the earth. Whether it's the trees in your neighborhood or the food on your plate, recognizing what you have can foster a deeper connection to the planet. This can help you develop a sense of responsibility towards the environment.

4. **Reflect on Your Impact**: Think about your daily habits. Are they helping or harming the environment? This reflection can be eye-opening. I once realized that my coffee habit was contributing to waste—so I switched to a reusable cup. Simple, right? But it made a difference. This self-awareness is crucial for identifying areas where you can make positive changes.

5. **Connect with Nature**: Spend time outdoors. Go for a hike, plant a garden, or just sit in a park. When you immerse yourself in nature, you'll feel that connection and want to protect it. Spending time in nature can reduce stress, improve mental well-being, and foster a deeper appreciation for the environment.

Now, let's chat about practices for reducing consumption and increasing awareness. It's easy to get caught up in the consumer culture we live in. Ads bombard us from every angle, telling us we need the latest gadget or trend. But what if we hit the brakes and asked ourselves, "Do I really need this?" This mindful approach to consumption can help us reduce waste and live more sustainably.

Here are some practical tips to help you cut back:

- **Adopt the 30-Day Rule**: Before making a purchase, wait 30 days. If you still want it after that time, then maybe it's worth it. More often than not, you'll find that the urge has passed. It's like a mental detox for your wallet. This helps to avoid impulse purchases and encourages more thoughtful consumption.

- **Declutter**: Go through your stuff. If you haven't used it in a year, consider donating or selling it. Not only does this free up space, but it also helps others. Win-win! This reduces waste and gives your unwanted items a second life.

- **Choose Quality Over Quantity**: Invest in fewer, high-quality items instead of a bunch of cheap stuff that'll fall apart in a month. Think of it as buying a good pair of boots that'll last you years instead of those trendy shoes you'll toss after a season. This reduces the need for frequent replacements and minimizes waste.

- **Mindful Eating**: This one's a biggie. When you eat, take your time. Savor each bite. This not only reduces food waste but also helps you appreciate what you're consuming. Plus, it's a great excuse to enjoy dessert a little longer! This can also help you make healthier food choices and improve your digestion.

- **Join a community**: Find like-minded folks who are also into sustainability. Whether it's a local group or an online community, surrounding yourself with others can keep you motivated. It's like having a support group for your eco-friendly journey. Connecting with others can provide support, inspiration, and valuable resources.

Now, you might be thinking, "But how does all this mindfulness stuff actually help the planet?" Great question! When you're mindful, you become more aware of your choices and their consequences. You start asking questions like, "Where does this come from?" and "What happens to it after I'm done?" That awareness can lead to ACTION, and action is what we need to create change. Mindfulness helps us connect the dots between our daily actions and their environmental impact.

Let me share a quick story. A few years back, I was in a coffee shop, and I overheard a couple of teens chatting about their plans to reduce their plastic use. They were pumped about it! They had created a social media challenge to encourage their friends to ditch single-use plastics. It hit me—these young folks were not just talking the talk; they were walking the walk. Their excitement was contagious, and it made me realize that every small action can ripple out into something bigger. This demonstrates the power of individual action and the potential for collective change.

So, what's the takeaway here? Mindfulness and sustainability are two sides of the same coin. When you practice mindfulness, you cultivate a deeper awareness of your impact on the planet. And when you embrace sustainability, you're making a conscious choice to live in harmony with the earth. It's a beautiful cycle, folks. These two concepts work together to create a more conscious and responsible way of living.

To wrap it up, here's a quick recap of actionable steps you can take:

- Start small and pick one sustainable habit to focus on.

- Educate yourself about sustainability through books, podcasts, and documentaries.

- Practice gratitude for the planet and reflect on your daily habits.

- Adopt the 30-day rule before making purchases and declutter your space.

- Choose quality over quantity and embrace mindful eating.

- Connect with a community of eco-conscious individuals.

Remember, every step you take matters. You don't have to be perfect; you just have to be willing to try. After all, the earth is our only home, and it deserves our love and care. By integrating mindfulness into our daily lives, we can create a more sustainable and harmonious relationship with the planet.

Chapter 15

Community Involvement

Have you ever watched a single pebble tossed into a calm pond? It sends ripples spreading outward in all directions, right? That's kind of what community action feels like, especially when it comes to environmental efforts. One person steps up, and before you know it, the whole neighborhood starts to buzz with a newfound energy. It's like when your friend shows up to a party in a wild outfit – everyone's curious, and suddenly there's a mini fashion revolution brewing right in your living room. Community involvement isn't just important; it's absolutely CRUCIAL if we want to build a sustainable future for ourselves and generations to come.

So, let's break it down a bit. Why should you even care? Well, here's the deal: community action amplifies the impact of what one person can do. When you team up with your pals, you can tackle those BIG problems that seem impossible to face alone. Imagine this: you and your friends decide to clean up your local park. You're not just picking up a few stray soda cans; you're setting an example for others. You're sparking conversations with fellow park-goers, and you're showing your neighbors that they can step up and make a difference too. Maybe you inspire them to bring their own reusable water bottles next time they visit the park.

Now, how do you get started? Or better yet, how do you hop on an existing initiative that aligns with your interests? Here's a little roadmap for you:

First off, do some digging! Research local initiatives. Check out social media platforms, local news websites, or community bulletin boards. Look for groups focused on sustainability, conservation, or climate action. Websites like Meetup.com can be a goldmine for finding folks who are just as passionate about the environment as you are.

Next, don't be shy – attend community meetings or local environmental workshops. Bring your enthusiasm and maybe a friend or two. You'll be surprised by how many people out there are just as fired up as you are about protecting the planet. Plus, it's a great way to meet new people who share your values.

Then, there's volunteering. Lots of organizations are always on the lookout for extra hands. Whether it's planting trees in a local park, organizing a clean-up event at the beach, or running awareness campaigns about recycling, there's definitely a spot for you. Volunteering is also a killer way to meet new friends who share your passion for the environment.

If you can't find what you're looking for, why not start your own initiative? Gather your friends, brainstorm some ideas, and take action! Maybe kick off a community garden in a vacant lot or start a recycling program at your school. Remember, it doesn't have to be a massive project to make a difference. Even small actions, like starting a composting bin at home, can inspire others to do the same.

Now, let's chat about building networks. Think of this as the glue that holds your community action together. When you connect with others who share your passion, you're not just sharing ideas; you're creating a collective impact that's way bigger than the sum of its parts. Here's how to do it:

Collaborate with local businesses! Many businesses are eager to support community initiatives, especially those that align with their sustainability goals. They might offer resources, sponsorship, or even encourage their employees to volunteer their time. Approach them with a clear plan, and you might be surprised at how quickly they jump on board.

Leverage the power of social media! Use platforms like Instagram, Facebook, or TikTok to spread the word about your initiatives. Create eye-catching events, share inspiring success stories, and use engaging captions to encourage others to get involved. The more people see your passion for the environment, the more likely they are to join your movement.

Create a supportive environment. Foster a culture of encouragement and celebrate small victories! When someone organizes a successful clean-up event, shout it from the rooftops (metaphorically speaking)! Recognition keeps folks motivated and engaged.

Network with other groups working towards similar goals. Connect with other environmental organizations in your area. Attend their events, collaborate on projects, and share resources. You'll not only expand your reach but also gain valuable insights from others who are experienced in the field.

But remember, community involvement isn't a solo gig. It's a team sport! The more you involve others, the more powerful your efforts become. Think of it like building a band – each person brings their own unique sound, and together, you create a symphony of change.

Let's not forget the importance of persistence. Change doesn't happen overnight. You might face setbacks, and that's totally okay. Keep pushing forward. Keep engaging your community. If you hit a wall, find a window. If there's no window, maybe it's time to build a door. Don't let challenges discourage you; view them as opportunities to learn and grow.

And here's a little pep talk for you: don't underestimate your ability to influence others. You might think you're just one person, but remember that every big movement started with someone saying, "Hey, we can do better." Your actions can inspire others to take action, too.

Now, let's wrap this up with some actionable steps. Here's what you can do to get involved in your community:

- Join or start a local environmental group. Find a few friends and brainstorm what environmental issues matter most to you in your local area.

- Volunteer for a local clean-up or tree-planting event. Check local community calendars or online resources; there's almost always something happening in your area.

- Reach out to local businesses for support. Draft a simple, polite email explaining your initiative and how they can help. You could even create a short presentation to share with them.

- Use social media to promote your efforts. Share your journey, post photos and videos of your activities, and encourage others to join in. Use relevant hashtags to reach a wider audience.

- Network, network, network! Attend local events and make connections with people who are passionate about sustainability.

So, what are you waiting for? Grab your friends, throw on your favorite eco-friendly shirt, and get out there. The planet needs YOU, and your community is waiting for your spark to ignite action. Let's create those ripples together!

Let's dive a bit deeper into why community involvement is so powerful. You see, when people come together, they create this buzz, this energy that can't be ignored. It's like a snowball effect. One action leads to another, and soon enough, you've got a whole avalanche of change rolling down the hill. It's not just about the immediate impact of the action itself, but also the long-term effects on community awareness and engagement.

Take the example of community gardens. They're not just about growing veggies. They're about bringing people together from diverse backgrounds, building relationships, and fostering a sense of belonging within the neighborhood. You plant a seed, both literally and metaphorically. You're cultivating not just crops but also community spirit. Plus, who doesn't love fresh, locally grown tomatoes?

And let's talk about the impact on mental health. Engaging in community action can significantly boost your mood. You're getting outside in the fresh air, meeting new people, and working towards something meaningful and positive. It's like a triple whammy of positivity. You're helping the environment, connecting with others, and feeling good about yourself all at once. That's a win-win-win if you ask me.

Now, let's throw in a little humor. You know how they say laughter is the best medicine? Well, it can also be a powerful tool in community action. Humor can break the ice, lighten the mood during challenging tasks, and make people more willing to engage. Imagine organizing a clean-up event with a funny theme—like "Trashy Fashion Day" where everyone dresses up in their most outrageous outfits made of recycled materials while picking up litter. You'll get people talking, laughing, and actually wanting to join in. Plus, you'll have some epic photos for social media!

And speaking of social media, it's a double-edged sword. On one hand, it's a fantastic platform to spread the word about your initiatives and rally support from a wide audience. On the other hand, it can sometimes be a breeding ground for negativity and misinformation. So, how do you navigate that? Focus on the positive. Share genuine success stories, highlight community heroes who are making a difference, and create a vibe that encourages people to join in rather than sit back and scroll passively.

Let's not forget about the younger generation. They're the future, right? Engaging kids and teens in community action is crucial. They bring fresh ideas, boundless energy, and a unique perspective. Plus, teaching them the importance of environmental stewardship early on can lead to lifelong habits. Think about organizing school events, like eco-friendly fairs with interactive games or educational workshops with hands-on activities. Make it fun and interactive—kids love hands-on activities.

But hey, it's not just about the young ones. Everyone has something valuable to contribute, regardless of age. You might find that your local retirees have a wealth of knowledge and experience to share, perhaps about local ecosystems or community history. Maybe they've been gardening for decades or have firsthand stories about how the community used to come together for different causes. Tap into that wisdom. It can be incredibly inspiring and provide valuable context.

Now, let's get a little philosophical for a sec. Community involvement is all about connection. In a world that often feels divided by various issues, coming together for a common cause like environmental protection can bridge gaps between people. It doesn't matter if you're from different backgrounds or have different opinions on other matters; when you're working side by side for a better environment, you start to see each other as allies instead of adversaries. It's a beautiful thing.

And let's be real—sometimes it's easy to get discouraged. You might feel like your efforts are just a drop in the bucket in the face of global environmental challenges. But think about this: every little bit counts. It's like trying to fill a swimming pool with a garden hose. It may take a while, but eventually, you'll see the water level rise. Your contributions, no matter how small they may seem at first, add up over time and contribute to a larger collective impact.

So, here's the deal. If you're feeling overwhelmed by the scale of environmental issues, just start small. Maybe pick one specific issue that speaks to you personally. Is it plastic waste in your neighborhood? The need for more urban green spaces? Wildlife conservation in your local area? Focus on that one thing, and then expand your efforts as you gain confidence, experience, and connections within your community. Remember, it's about progress, not perfection.

And don't forget to celebrate those victories, no matter how small they may seem. Did you manage to get a handful of people to show up for a clean-up event? Awesome! Post about it on social media. Share the pictures and videos. Let others see the joy and camaraderie that comes from working together towards a common goal. It'll inspire them to join in next time and create a positive feedback loop.

Now, let's talk about the nitty-gritty of organizing events. Planning can sometimes feel daunting, but it doesn't have to be. Start with a clear and concise goal. What do you want to achieve with this event? Then, break it down into smaller, more manageable steps. Create a timeline with deadlines for each task, delegate tasks among your team members, and don't be afraid to ask for help from experienced organizers or community members. You'd be surprised how many people are willing to pitch in if you just ask.

And always keep the lines of communication open among your team and with the community. Whether it's through group chats, regular email updates, or engaging posts on social media, make sure everyone's on the same page. Share updates on your progress, celebrate milestones together, and keep the enthusiasm flowing to maintain momentum.

But here's a little secret: not every event will go perfectly according to plan. You might have lower turnout than expected or face unexpected logistical challenges. And that's okay! Roll with the punches. Learn from each experience, and don't let it discourage you from future efforts. Remember, it's all part of the journey.

Now, let's wrap it up with some final thoughts. Community involvement is a truly powerful force for positive change. It's about connection, meaningful collaboration, and making a tangible difference in the world around you. So, whether you're joining an existing initiative or taking the lead and starting your own, just remember this: you've got the power to create ripples of positive change that extend far beyond what you can imagine.

So, grab your friends, put on your favorite eco-friendly shirt, and get out there. The planet needs YOU, and your community is waiting for your spark to ignite action. Let's create those ripples together!

Chapter 16

The Future of Our Planet

Hey there, friends! Let's chat about something that's been buzzing in my brain – a sustainable future. You know, the kind where we don't just push our problems onto the next generation, leaving them to deal with the consequences. Imagine this: fresh, clean air filling your lungs, vibrant ecosystems teeming with life, and a planet that's not wheezing for survival. Sounds pretty sweet, huh? Well, guess what? You're a crucial part of making this dream a reality. Yup, you! Don't roll your eyes at me; I'm dead serious! Your actions, even small ones, can make a huge difference.

Thinking about a sustainable future is like creating a masterpiece. It takes a wild imagination, sure, but it also needs some serious commitment and practical steps. Back when I was a teenager, I'd doodle on the back of my math homework when I was supposed to be solving equations. I'd sketch a world where everyone recycled their trash, solar panels dotted the rooftops of houses and buildings, and bikes were the norm for getting around instead of gas-guzzling cars. I was just a kid with a wild imagination, but those doodles lit a fire in me. I didn't just want to dream about a better future; I wanted to make a difference in the real world. So, let's break it down. What does this sustainable future actually look like?

Picture a clean environment. Imagine strolling through a local park bursting with tall, leafy trees, colorful flowers blooming in vibrant hues, and the sweet sound of birds chirping their cheerful songs. That's not just some fairy tale or a scene from a movie; it's a reality we can create if we all pitch in and do our part.

Next up, renewable energy. Think solar panels soaking up the sun's rays on rooftops, converting that sunlight into clean electricity, and wind turbines twirling gracefully in the breeze, generating power without polluting the air. This isn't just for the hardcore eco-warriors; it's for everyone. Seriously, it's like having a buffet of energy options—sun, wind, water, and more—and we're all invited to partake.

Then there's community engagement. Imagine people gathering in their neighborhoods to plant trees along city streets, clean up local rivers and streams, and push for policies that protect our planet for future generations. It's like a neighborhood potluck, but instead of bringing casseroles and desserts, we're sharing ideas and taking concrete actions to make our world a better place.

So, how do we get to this beautiful future? It all kicks off with education and awareness. Knowledge is power, right? Think about it. When I first learned about climate change and its potential consequences, it felt like a light bulb flicked on in my brain. I had no clue how much our everyday choices, like driving a car or using disposable plastic, impacted the planet.

Education isn't just for classrooms and textbooks; it's everywhere you look. Social media platforms, informative documentaries, engaging podcasts—these are our modern-day tools for learning and spreading awareness. Use them! Follow eco-conscious influencers who are sharing helpful tips, binge-watch films that spotlight important environmental issues, and dive into books that challenge your thinking and offer new perspectives. And hey, don't keep that knowledge to yourself! Share what you learn with your friends and family. It spreads like wildfire when you start chatting with the people closest to you.

But here's the kicker: awareness without action is like a car running on empty. You can have the best intentions in the world, but if you're not moving forward and taking concrete steps, you're just sitting there, going nowhere. So, what are some real, practical steps we can take to make a difference?

Start small. Seriously, don't think you have to save the entire planet overnight. Pick one small change to focus on implementing in your daily life. Maybe it's switching to a reusable water bottle instead of buying those pesky plastic ones that end up in landfills. Easy peasy, right?

Get involved in your local community! Join a local environmental group or volunteer for clean-up events in your area. I once jumped into a river clean-up with some friends, and let me tell you, it was a blast! I met some awesome folks who cared about the environment, got some fresh air and exercise, and felt like I was part of something way bigger than myself.

Advocate for change! Speak up about the issues that matter to you! Write letters or emails to your local representatives about environmental concerns in your community. If you don't tell them what you care about, how will they know what's important to you?

Educate others in your community. Be the change-maker in your own circle of friends and family. Host a movie night featuring thought-provoking documentaries about climate change or sustainable living practices. Who knows? You might just spark someone else's passion for protecting the planet.

And let's not forget about the future generations. We owe it to them to leave behind a world that's not just livable but thriving and healthy. Think of it like planting a tree in your backyard. You might not see it grow to its full height and provide shade for decades, but you're laying the groundwork for future generations to enjoy its benefits.

Here's a little challenge for you: take a moment to think about your personal goals when it comes to sustainability.

What specific actions do you want to take? What impact do you want to have? Maybe it's cutting down your personal carbon footprint or advocating for cleaner energy sources in your community. Write it down! Making it real on paper can help you stay focused and motivated.

And don't sweat it if you stumble along the way. Trust me, I've tripped over my own two feet more times than I can count when trying to live a more sustainable lifestyle. It's all part of the journey of learning and growing. What matters most is that you keep pushing forward and continue to make an effort.

Now, let's talk numbers for a sec. Did you know that if every American swapped just one meal a week for a plant-based one, it'd be like taking 7.6 million cars off the road for an entire year? That's a stat that hits hard and shows the power of collective action! It's these small, individual changes that can stack up to make a HUGE positive impact on the environment.

In this tech-driven world, we've got access to some amazing tools that can help us stay informed and engaged in sustainability efforts. Apps like "Oroeco" help you track your personal carbon footprint, showing you the impact of your daily choices, while "Good On You" rates fashion brands based on their sustainability practices, helping you make more ethical purchasing decisions. Use these resources to guide your choices and make informed decisions.

And here's a fun metaphor for you: think of our planet as a giant puzzle. Each of us holds a piece of that puzzle, and if we don't work together to put it all together, we'll never see the full, beautiful picture. So, let's collaborate with each other, share our ideas and experiences, and take collective action to build that sustainable future.

To wrap it up, envisioning a sustainable future isn't just some lofty, idealistic dream; it's a shared responsibility that we all have. Education and awareness are the fuel that ignites the change, and consistent action is the powerful engine that drives us forward. Let's commit to taking those actionable steps, no matter how small or insignificant they may seem at first.

Remember, you're not alone in this important effort. We're all in it together, working towards a common goal. So, what are you waiting for? Let's roll up our sleeves, get to work, and start creating a brighter, more sustainable future for ourselves and for generations to come. The planet's counting on us.

Now, go out there and make a difference. The future's in your hands, and trust me, it's worth fighting for!

But let's not just stop there. Think about how you can weave sustainability into your everyday life. It's not just about the big, dramatic actions; it's the small, consistent choices that pile up and create real change over time. Like, how often do you grab takeout food? Maybe try cooking at home more often. Not only is it usually healthier for you, but it also significantly cuts down on all that unnecessary packaging waste. Plus, you get to show off your culinary skills (or at least try to!), and maybe even discover a new favorite dish.

And what about your wardrobe? Fast fashion is a huge contributor to environmental waste and pollution. Why not hit up a local thrift store or organize a clothing swap with your friends? It's like a treasure hunt! You'd be surprised at what stylish gems you can find at a fraction of the price, and it's way more fun and rewarding than just mindlessly scrolling through online shopping sites.

Let's chat about transportation for a sec. If you can, ditch the car for short trips around your neighborhood. Walk, ride your bike, or hop on public transport if it's available in your area. Not only does this significantly reduce harmful emissions, but it's also a great way to explore your neighborhood and get some exercise. You might discover a cute little coffee shop you didn't even know existed or a hidden park with a beautiful view!

And speaking of coffee, how many of us grab that much-needed morning cup in a disposable cup? It's a habit many of us have, but it contributes to a lot of waste. Invest in a reusable travel mug or coffee cup. It's a small change that can make a big difference over time. Plus, some coffee shops even give you a small discount for bringing your own cup, which is a win-win!

You know what else is super impactful for local economies and the environment? Supporting local businesses in your community. When you choose to shop local, you're not just getting fresh produce or unique, handcrafted items; you're also significantly reducing the carbon footprint associated with transporting goods over long distances. It's a double whammy of positive impact for the planet and your local community!

Now, let's not forget about the issue of waste. Get into composting! It's easier than you might think. You can turn your kitchen scraps, like fruit and vegetable peels, into nutrient-rich soil for your garden. It's like giving back to the earth in a tangible way while simultaneously reducing the amount of waste that ends up in landfills. Talk about a win-win!

And if you're feeling extra ambitious and have some space, why not start a community garden in your neighborhood? It's a fantastic way to bring people together, grow your own fresh, healthy food, and promote biodiversity in your local ecosystem. Plus, it's a great excuse to get your hands dirty and connect with nature!

Oh, and let's not forget about water conservation. We often take clean water for granted, but it's a precious and limited resource. Simple things like fixing leaky faucets and toilets, taking shorter showers, or using a broom instead of a water hose to clean your driveways and sidewalks can make a significant difference in water usage. Every drop counts, right?

And let's talk about the power of social media for good. Use your online platforms to spread awareness about sustainability and inspire others to take action. Share helpful tips, highlight local environmental events and initiatives, or even just post about your own sustainable choices and experiences. You'd be amazed at how many people you can reach and inspire just by sharing your own journey.

But let's keep it real and acknowledge that change doesn't happen overnight. It's a marathon, not a sprint, when it comes to creating a sustainable future. There'll inevitably be bumps and challenges along the way, and that's perfectly okay. What's truly important is that we keep pushing forward, keep learning from our mistakes, and continue to grow and improve our efforts.

So, take a moment to reflect on your own life. What's one small, manageable change you can make today to live more sustainably? Maybe it's as simple as remembering to carry a reusable shopping bag with you to the grocery store or choosing to walk or bike instead of driving your car for short distances. Whatever it is, make it count.

In the end, it's all about building a culture of sustainability in our communities and beyond. It's about making sustainable practices the norm, not the exception. And that transformation starts with each and every one of us, making conscious choices in our daily lives. Together, we can create a powerful ripple effect that leads to real and lasting positive change for our planet.

The future is bright and full of potential, but it's up to us to keep it that way by taking action now. The planet's counting on us, and trust me, it's worth every single effort we make.

Now, go out there and be the change you want to see in the world. The future's in your hands, and it's time to make it happen!

Chapter 17

Final Thoughts

Thank you for taking the time to read this book. I hope it has inspired you, offered clarity, and provided tools to support your journey toward personal growth and fulfillment. Every step forward, no matter how small, is a testament to your commitment to a better, brighter future. Each page turned, each concept considered, represents a step on your path.

If this book resonated with you or made a positive impact on your life, I would deeply appreciate it if you could take a moment to share your thoughts by leaving a review. Your feedback not only helps others discover this book but also encourages me to continue writing and sharing ideas that foster growth and change. Your honest opinion, whether positive or constructive, is invaluable.

Remember, your story is still unfolding, and every chapter is an opportunity to shape the life you want to lead. Keep moving forward with courage and intention. Embrace the challenges, celebrate the victories, and continue writing your own unique story. You are the author of your own life, and you have the power to create a fulfilling and meaningful narrative.

Final Thoughts

Thank you for being part of this journey. Wishing you all the best on yours. May your path be filled with joy, growth, and fulfillment.